AF754814

Introduction to Time Series Analysis and Forecasting

Introduction to Time Series Analysis and Forecasting

Editor

Lavra Filipek

scitus
academics

Introduction to Time Series Analysis and Forecasting
Edited by **Lavra Filipek**

ISBN: 978-1-68117-191-3

Library of Congress Control Number: 2015949135

© 2016 by
SCITUS Academics LLC,
www.scitusacademics.com
Box No. 4766, 616 Corporate Way,
Suite 2, Valley Cottage,
NY 10989

This book contains information obtained from highly regarded resources. Copyright for individual articles remains with the authors as indicated. All chapters are distributed under the terms of the Creative Commons Attribution License, which permits unrestricted use, distribution, and reproduction in any medium, provided the original author and source are credited.

Notice

Reasonable efforts have been made to publish reliable data and views articulated in the chapters are those of the individual contributors, and not necessarily those of the editors or publishers. Editors or publishers are not responsible for the accuracy of the information in the published chapters or consequences of their use. The publisher believes no responsibility for any damage or grievance to the persons or property arising out of the use of any materials, instructions, methods or thoughts in the book. The editors and the publisher have attempted to trace the copyright holders of all material reproduced in this publication and apologize to copyright holders if permission has not been obtained. If any copyright holder has not been acknowledged, please write to us so we may rectify.

Printed in United States of America on Acid Free Paper ∞

Preface

A time series is a collection of data recorded over a period of time—weekly, monthly, quarterly, or yearly. Forecasting the level of sales, both short-term and long-term, is practically dictated by the very nature of business organizations. Competition for the consumer's dollar, stress on earning a profit for the stockholders, a desire to procure a larger share of the market, and the ambitions of executives are some of the prime motivating forces in business. Thus, a forecast is necessary to have the raw materials, production facilities, and staff available to meet the projected demand.

Time series analysis can be applied to real-valued, continuous data, discrete numeric data, or discrete symbolic. Analyzing time oriented data and forecasting future values of a time series are among the most important problems that analysis face in many fields ranging from finance and economics to managing production operations. The emphasis of this book is on time series analysis and forecasting. This book is intended for practitioners who make real world forecasts. Time series analysis has got attention of many researches from different fields, such as business administration, economics, public finances.

Forecasting is an important activity in economics, commerce, marketing and various branches of science. This book, Introduction to Time Series Analysis and Forecasting, is concerned with forecasting methods based on the use of time-series analysis. It is primarily intended as a reference source for practitioners and researchers in forecasting, who could, for example, be statisticians, econometricians, operational researchers, management scientists or decision scientists.

Contents

Condition For Successful Square Transformation In Time Series Modeling

J. Ohakwe[1], *O. Iwuoha*[2], *E. L. Otuonye*[2]

[1]Department of Statistics, Faculty of Science, Federal University Otuoke, Yenagoa, Nigeria

[2]Department of Statistics, Faculty of Biological and Physical Sciences, Abia State University, Uturu, Nigeria

1

ABSTRACT

In this study we establish the probability density function of the square transformed left-truncated $N(1,\sigma^2)$ error component of the multiplicative time series model and the functional expressions for its mean and variance. Furthermore the mean and variance of the square transformed left-truncated $N(1,\sigma^2)$ error component and those of the untransformed component were compared for the purpose of establishing the interval for $\boldsymbol{\sigma}$ where the properties of the two distributions are approximately the same in terms of equality of means and normality. From the results of the study, it was established that the two distributions are normally distributed and have means $\cong 1.0$ correct to 1 dp in the interval $0 < \boldsymbol{\sigma} < 0.027$, hence a successful square transformation where necessary is achieved for values of $\boldsymbol{\sigma}$ such that $0 < \boldsymbol{\sigma} < 0.027$.

INTRODUCTION

Consider a normally distributed random variable X with probability density function f(x) specified as

$$f(x) = \frac{1}{\sigma\sqrt{2\pi}} e^{-\frac{1}{2}\left(\frac{x-1}{\sigma}\right)^2}, -\infty < x < \infty, \sigma^2 > 0 \tag{1}$$

Often in practice, the random variable X which has a $N(1,\sigma^2)$ distribution do not admit values less than or equal to zero. We therefore disregard or truncate all values of $X \le 0$ to take care of the admissible region X > 0. Now if the values of X below or equal to zero cannot be observed due to censoring or truncation, then the resulting distribution is a left-truncated normal distribution.

[1] obtained the probability density function of the left-truncated normal distribution as

$$f(x) = \frac{e^{-\frac{1}{2}\left(\frac{x-1}{\sigma}\right)^2}}{\sigma\sqrt{2\pi}\left[1-\Phi\left(-\frac{1}{\sigma}\right)\right]}, 0 < x < \infty \tag{2}$$

with mean (E(X)) and variance (Var(X)) given by

$$E(X) = 1 + \frac{\sigma e^{\frac{-1}{2\sigma^2}}}{\sqrt{2\pi}\left[1-\Phi\left(\frac{-1}{\sigma}\right)\right]} \tag{3}$$

And

$$\mathrm{Var}(X) = \frac{\sigma^2}{2\left[1-\Phi\left(\frac{-1}{\sigma}\right)\right]}\left[1+\Pr\left[\chi^2_{(1)} < \frac{1}{\sigma^2}\right]\right] - \frac{\sigma e^{-\frac{1}{2\sigma^2}}}{\sqrt{2\pi\left[1-\Phi(-1/\sigma)\right]}} - \left[\frac{\sigma e^{-\frac{1}{2\sigma^2}}}{\sqrt{2\pi\left[1-\Phi\left(-\frac{1}{\sigma}\right)\right]}}\right]^2 \tag{4}$$

The study of the properties of normally distributed random variables when certain outcomes are constrained or restricted has been a rich and fertile one with applications in regression analysis, inventory management and time series modeling to mention but a few.

A time series is a collection of observations made sequentially in time. Examples occur in a variety of fields, ranging from economics to engineering and methods of analyzing time series constitute an important area of statistics [2]. Time series analysis comprises methods that attempt to understand such time series, often either to understand the underlying context of the data points (Where did they come from? What generated them?), or to make forecasts. Time series forecasting is the use of a model to forecast or predict future events based on known past events.

Methods for time series analyses are often divided into three classes: descriptive methods, time domain methods and frequency domain methods. Frequency domain methods centre on spectral analysis and recently wavelet analysis [3,4] can be regarded as model-free analyses. Time domain methods [5,6] have a distribution-free subset consisting of the examination of the autocorrelation and cross-correlation analysis.

Descriptive methods [2,7] involve the separation of an observed time series into components representing trend (long term direction), the seasonal (systematic, calendar related movements), cyclical (long term oscillations or swings about the trend) and irregular (unsystematic, short term fluctuations) components. The descriptive method is known as time series decomposition. If short period of time are involved, the cyclical component is superimposed into the trend [2] and the observed time series $(X_t, t = 1, 2, \ldots, n)$ can be decomposed into the trend-cycle component (M_t), seasonal component (S_t) and the irregular/residual component (e_t).

The decomposition model of interest in this study is the multiplicative time series model given by

$$X_t = M_t S_t e_t \tag{5}$$

where e_t are independent, identically distributed normal errors with mean 1 and variance $\sigma^2 > 0 \left(i.e.\ e_t \sim N\left(1, \sigma^2\right)\right)$.

Data transformation is a mathematical operation that changes the measurement scale of a variable. Reasons for transformation include stabilizing variance, normalizing, reducing the effect of outliers, making a measurement scale more meaningful, and to linearize a relationship [8]. For further details on reasons for transformation, see [9- 11]. Many time series analyst assume normality and it is well known that variance stabilization implies normality of the series. The most popular and common data transformations are the power transformation namely the logarithm, square root, inverse, inverse square root, square and inverse square transformations. A statistical procedure for choice of appropriate data transformation can be obtained in [8,11]. It is important to note that, if we apply square transformation on model (5), we still obtain a multiplicative time series model given by

$$Y_t = M_t^2 S_t^2 e_t^2 = M_t^* S_t^* e_t^* \tag{5b}$$

where $M_t^2 = M_t^*, S_t^2 = S_t^*$ and $e_t^2 = e_t^*$

Studies on the effects of transformation on the error component of the multiplicative time series model) are not new in the statistical literature. The overall aim of such studies is to establish the conditions for successful transformation. A successful transformation is achieved when the desirable properties of a data set remains unchanged after transformation. The basic properties or assumptions of interest for this study are: 1) Unit mean and 2) constant variance. In this end, [1] investigated the effect of logarithmic transformation on the error component (e_t) of a multiplicative time series model where $\left(e_t \sim N\left(1, \sigma^2\right)\right)$ and discovered that the logarithm transform; $Y = \text{Log}\ e_t$ can be assumed to be normally distributed with mean, zero and the same variance, s^2 for $s < 0.1$. Similarly [12,13] had studied the effects of inverse and square root transformations on the error component of the same model. [12] discovered that the inverse transform $Y = \frac{1}{e_t}$ can be assumed to be normally distributed with mean, one and the same variance provided $\sigma \leq 0.07$. Similarly [13] discovered that the square root transform; $Y = \sqrt{e_t}$ can be assumed to be normally distributed with unit mean and variance, $4\sigma^2$ for $\sigma \leq 0.3$, where σ^2 is the variance of the original error component before transformation. Furthermore, [14] has studied the implication of square root transformation on a two-parameter Gamma distributed error component of a multiplicative error model and discovered that the unit mean assumption is approximately maintained, but the variance of the transformed distributions is one-quarter of the original variance.

In this paper we study the implication of square transformation on the error component of the multiplicative time series model with a view to establish the interval for σ, for which the transformation is successful. The paper is organized into 6 sections. Section 1 contains the introduction. The probability density function, mean and variance of the square transformed left-truncated $N(1,\sigma^2)$ error component are established in Section 2. Comparison of the square transformed and the untransformed distributions were compared in Section 3. Finally the summary and conclusion, references and appendix are respectively contained in Sections 4-6.

PROBABILITY DISTRIBUTION OF THE SQUARE TRANSFORMED ERROR COMPONENT

Using the transformation,

$$y = x^2 \tag{6}$$

in (2) and the admissible values of $y(0 < y < \infty)$, we would then find the probability density function (pdf) of $y=x^2$. From this point forward it is important to note from (5) and (5b), that symbolically $y = e_t^* = e_t^2$ and $x = e_t$. Applying the transformation in (6) implies that

$$\frac{dx}{dy} = \frac{1}{2\sqrt{y}} \tag{7}$$

But, the pdf of y, f(y) is given by

$$f(y) = f\left(x = \sqrt{y}\right)\left|\frac{dx}{dy}\right| \tag{8}$$

where $\left|\frac{dx}{dy}\right| = \frac{1}{2\sqrt{y}}$ is the absolute value of the Jacobian of the transformation [15]. Thus

$$f(y) = \frac{1}{\sigma\sqrt{2\pi}\left(1-\Phi\left(-\frac{1}{\sigma}\right)\right)} e^{-\frac{1}{2\sigma^2}\left(\sqrt{y}-1\right)^2}\left|\frac{dx}{dy}\right|$$

$$= \frac{1}{\sigma\sqrt{2\pi}\left(1-\Phi\left(-\frac{1}{\sigma}\right)\right)} e^{-\frac{1}{2\sigma^2}\left(\sqrt{y}-1\right)^2}\frac{1}{2\sqrt{y}}$$

hence

$$f(y) = \begin{cases} \dfrac{e^{-\frac{1}{2\sigma^2}\left(\sqrt{y}-1\right)^2}}{2\sigma\sqrt{2\pi}\left(1-\Phi\left(-\frac{1}{\sigma}\right)\right)\sqrt{y}}, & 0 < y < \infty \\ 0, & \text{elsewhere} \end{cases} \tag{9}$$

The crucial question is now "is (9) a proper pdf?". If it is to be a proper pdf, it must satisfy the condition;

$$\int_0^\infty f(y)\,\mathrm{d}y = 1$$

1)

hence we now proceed to show that the integral of (9) is equal to unity as follows:

$$\int_0^\infty f(y)\,\mathrm{d}y = \int_0^\infty \frac{e^{\frac{-1}{2}\left(\frac{\sqrt{y}-1}{\sigma}\right)^2}}{2\sigma\sqrt{2\pi}\left(1-\Phi\left(\frac{-1}{\sigma}\right)\right)\sqrt{y}}\mathrm{d}y \tag{10}$$

Let

$$u = \frac{\sqrt{y}-1}{\sigma} \tag{11a}$$

hence

$$-\frac{1}{\sigma} < u < \infty, y = (\sigma u + 1)^2 \tag{11b}$$

and

$$\mathrm{d}y = 2\sigma(\sigma u + 1)\,\mathrm{d}u \tag{11c}$$

therefore, substituting the results in (11a) through (11c) into (10) yields

$$\int_0^\infty f(y)\,\mathrm{d}y = \frac{1}{\sqrt{2\pi}\left(1-\Phi\left(\frac{-1}{\sigma}\right)\right)}\int_{-1/\sigma}^{\infty} \mathrm{e}^{-\frac{1}{2}u^2}\,\mathrm{d}u$$

$$= \frac{1}{\left(1-\Phi\left(\frac{-1}{\sigma}\right)\right)}\int_{-1/\sigma}^{\infty} \frac{1}{\sqrt{2\pi}}\mathrm{e}^{-\frac{1}{2}u^2}\,\mathrm{d}u$$

$$= \frac{1}{\left(1-\Phi\left(\frac{-1}{\sigma}\right)\right)} P\left(u \geq -\frac{1}{\sigma}\right)$$

$$= \frac{\left(1-P\left(u < -\frac{1}{\sigma}\right)\right)}{\left(1-\Phi\left(-\frac{1}{\sigma}\right)\right)} = \frac{\left(1-\Phi\left(\frac{-1}{\sigma}\right)\right)}{\left(1-\Phi\left(\frac{-1}{\sigma}\right)\right)} = 1$$

and this shows that (9) is a proper pdf.

Mean of the Square Transformed Distribution, $\mathrm{E}(\mathrm{Y})\left(\mathrm{Y}=\mathrm{X}^2\right)$

By definition

$$E(Y) = \int_0^\infty y f(y)\,\mathrm{d}y = \int_0^\infty \frac{y\mathrm{e}^{\frac{-1}{2}\left(\frac{\sqrt{y}-1}{\sigma}\right)^2}}{2\sigma\sqrt{2\pi}\left(1-\Phi\left(\frac{-1}{\sigma}\right)\right)\sqrt{y}}\,\mathrm{d}y$$

$$= \frac{1}{2k\sigma}\int_0^\infty y^{\frac{1}{2}}\mathrm{e}^{\frac{-1}{2}\left(\frac{\sqrt{y}-1}{\sigma}\right)^2}\,\mathrm{d}y \qquad (12)$$

where

$$k = \sqrt{2\pi}\left(1-\Phi\left(\frac{-1}{\sigma}\right)\right).$$

Applying the substitution given in (11) into (12), we obtain

$$E(Y)=\frac{2\sigma}{2k\sigma}\int_{-1/\sigma}^{\infty}(\sigma\mu+1)^2 e^{-\frac{u^2}{2}}du$$

$$=\frac{1}{k}\int_{-1/\sigma}^{\infty}(\sigma\mu+1)^2 e^{-\frac{u^2}{2}}du=\frac{1}{k}\int_{-1/\sigma}^{\infty}\left(1+\sigma\mu+\sigma^2u^2\right)e^{-\frac{u^2}{2}}du$$

$$=\frac{1}{k}\left[\int_{-1/\sigma}^{0}e^{-\frac{u^2}{2}}du+2\sigma\int_{-1/\sigma}^{0}ue^{-\frac{u^2}{2}}du+\sigma^2\int_{-1/\sigma}^{0}u^2e^{-\frac{u^2}{2}}du\right]$$

$$=\frac{1}{k}\left[\int_{-1/\sigma}^{0}e^{-\frac{u^2}{2}}du+2\sigma\int_{-1/\sigma}^{0}ue^{-\frac{u^2}{2}}du+\sigma^2\int_{-1/\sigma}^{0}u^2e^{-\frac{u^2}{2}}du\right]$$

$$+\frac{1}{k}\left[\int_{0}^{1/2\sigma^2}e^{-\frac{u^2}{2}}du+2\sigma\int_{0}^{1/2\sigma^2}ue^{-\frac{u^2}{2}}du+\sigma^2\int_{0}^{1/2\sigma^2}u^2e^{-\frac{u^2}{2}}du\right]$$

$$=\frac{1}{k}\left[\psi(u)+\theta(u)\right] \tag{13}$$

where

$$\psi(u)=\int_{-1/\sigma}^{0}e^{-\frac{u^2}{2}}du+2\sigma\int_{-1/\sigma}^{0}ue^{-\frac{u^2}{2}}du+\sigma^2\int_{-1/\sigma}^{0}u^2e^{-\frac{u^2}{2}}du$$

and

$$\theta(u)=\int_{0}^{1/2\sigma^2}e^{-\frac{u^2}{2}}du+2\sigma\int_{0}^{1/2\sigma^2}ue^{-\frac{u^2}{2}}du+\sigma^2\int_{0}^{1/2\sigma^2}u^2e^{-\frac{u^2}{2}}du$$

If we let

$$p = \frac{u^2}{2}, -\frac{1}{\sigma} < u < 0 \tag{14}$$

In $\psi(u)$, we obtain

$$u^2 = 2p; u = 2^{\frac{1}{2}} p^{\frac{1}{2}}; 0 < p < \frac{1}{2\sigma^2} \tag{15a}$$

and

$$\mathrm{d}u = \frac{p^{-\frac{1}{2}}\mathrm{d}p}{\sqrt{2}} \tag{15b}$$

hence

$$\psi(u) = \frac{1}{\sqrt{2}} \int_0^{1/2\sigma^2} p^{-\frac{1}{2}} \mathrm{e}^{-p} \mathrm{d}p + \sigma \int_0^{1/2\sigma^2} \mathrm{e}^{-p} \mathrm{d}p + \sigma^2 \sqrt{2} \int_0^{1/2\sigma^2} p^{\frac{1}{2}} \mathrm{e}^{-p} \mathrm{d}p \tag{16}$$

By integration by parts the following results are obtained

$$\int_0^{1/2\sigma^2} p^{\frac{3}{2}} \mathrm{e}^{-p} \mathrm{d}p = \frac{3}{2}\sqrt{\pi}\Phi\left(\sqrt{\frac{1}{2\sigma^2}}\right) - \frac{3}{2}\mathrm{e}^{-\frac{1}{2\sigma^2}}\sqrt{\frac{1}{2\sigma^2}} - \frac{1}{2\sigma^2\sqrt{2\sigma^2}}\mathrm{e}^{-\frac{1}{2\sigma^2}} \tag{17a}$$

$$\int_{0}^{1/2\sigma^2} p e^{-p} dp = 1 - e^{-\frac{1}{2\sigma^2}} - \frac{1}{2\sigma^2} e^{\frac{1}{2\sigma^2}} \tag{17b}$$

$$\int_{0}^{1/2\sigma^2} p^{\frac{1}{2}} e^{-p} dp = \frac{1}{2}\sqrt{\pi}\Phi\left(\sqrt{\frac{1}{2\sigma^2}}\right) - \sqrt{\frac{1}{2\sigma^2}} e^{-\frac{1}{2\sigma^2}} \tag{17c}$$

$$\int_{0}^{1/2\sigma^2} e^{-p} dp = 1 - e^{-\frac{1}{2\sigma^2}} \tag{17d}$$

$$\int_{0}^{1/2\sigma^2} p^{-\frac{1}{2}} e^{-p} dp = \sqrt{\pi}\Phi\left(\sqrt{\frac{1}{2\sigma^2}}\right) \tag{17e}$$

hence

$$\psi(u) = \sigma + \left(1+\sigma^2\right)\sqrt{\frac{\pi}{2}}\Phi\left(\sqrt{\frac{1}{2\sigma^2}}\right) - 2\sigma e^{-\frac{1}{2\sigma^2}} \tag{18}$$

Furthermore, if we let

$$p = \frac{u^2}{2}, 0 < u < \infty \tag{19}$$

In $\theta(u)$, we obtain

$$u^2 = 2p; u = 2^{\frac{1}{2}} p^{\frac{1}{2}}; 0 < p < \infty \tag{20a}$$

and

$$du = \frac{p^{-\frac{1}{2}}dp}{\sqrt{2}} \tag{20b}$$

thus

$$\theta(u) = \frac{1}{\sqrt{2}}\int_0^{\infty} p^{-\frac{1}{2}}e^{-p}dp + \sigma\int_0^{\infty} e^{-p}dp + \sigma^2\sqrt{2}\int_0^{\infty} p^{\frac{1}{2}}e^{-p}dp$$
$$= \sigma + \sqrt{\frac{\pi}{2}}\left[1+\sigma^2\right] \tag{21}$$

Substituting the results of (18), (21) and the value of k into (13), we obtain

$$E(Y) = \frac{\left[2\sigma\left(1-e^{-\frac{1}{2\sigma^2}}\right)+\sqrt{\frac{\pi}{2}}\left(1+\sigma^2\right)\left(1+\Phi\left(\sqrt{\frac{1}{2\sigma^2}}\right)\right)\right]}{\sqrt{2\pi}\left(1-\Phi\left(-\frac{1}{\sigma}\right)\right)} \tag{22}$$

Variance of the Square Transformed Distribution $\left(Y = X^2\right), Var(Y)$

By definition

$Var(Y) = E\left(Y^2\right) - \left(E(Y)\right)^2$ but

$$E\left(Y^2\right)=\int_0^\infty y^2 f(y)\,\mathrm{d}y$$

$$=\int_0^\infty \frac{y^2 e^{\frac{-1}{2}\left(\frac{\sqrt{y}-1}{\sigma}\right)^2}}{2\sigma\sqrt{2\pi}\left(1-\Phi\left(\frac{-1}{\sigma}\right)\right)\sqrt{y}}\,\mathrm{d}y$$

$$=\frac{1}{2k\sigma}\int_0^\infty y^{\frac{3}{2}} e^{\frac{-1}{2}\left(\frac{\sqrt{y}-1}{\sigma}\right)^2}\,\mathrm{d}y \tag{23}$$

Applying the transformation given in (11) into (23), we have that

$$E\left(Y^2\right)=\frac{1}{k}\int_{-1/\sigma}^{\infty}(\sigma u+1)^4 e^{-\frac{u^2}{2}}\,\mathrm{d}u$$

$$=\frac{1}{k}\int_{-1/\sigma}^{0}(\sigma u+1)^4 e^{-\frac{u^2}{2}}\,\mathrm{d}u+\frac{1}{k}\int_0^\infty(\sigma u+1)^4 e^{-\frac{u^2}{2}}\,\mathrm{d}u \qquad 24)$$

But

$$\int_{-1/\sigma}^{0}(\sigma u+1)^4 e^{-\frac{u^2}{2}}\,\mathrm{d}u$$

$$=\sigma^4\int_{-1/\sigma}^{0}u^4 e^{-\frac{u^2}{2}}\,\mathrm{d}u+4\sigma^3\int_{-1/\sigma}^{0}u^3 e^{-\frac{u^2}{2}}\,\mathrm{d}u$$

$$+6\sigma^2\int_{-1/\sigma}^{0}u^2 e^{-\frac{u^2}{2}}\,\mathrm{d}u+4\sigma\int_{-1/\sigma}^{0}u e^{-\frac{u^2}{2}}\,\mathrm{d}u+\int_{-1/\sigma}^{0}e^{-\frac{u^2}{2}}\,\mathrm{d}u \tag{25}$$

and

$$\int_0^\infty (\sigma u+1)^4 e^{-\frac{u^2}{2}} du$$

$$= \sigma^4 \int_0^\infty u^4 e^{-\frac{u^2}{2}} du + 4\sigma^3 \int_0^\infty u^3 e^{-\frac{u^2}{2}} du$$

$$+ 6\sigma^2 \int_0^\infty u^2 e^{-\frac{u^2}{2}} du + 4\sigma \int_0^\infty u e^{-\frac{u^2}{2}} du + \int_0^\infty e^{-\frac{u^2}{2}} du \tag{26}$$

Applying the substitution in (14) and its corresponding results in (15) into (26), we have that

$$\sigma^4 \int_{-1/\sigma}^{0} u^4 e^{-\frac{u^2}{2}} du + 4\sigma^3 \int_{-1/\sigma}^{0} u^3 e^{-\frac{u^2}{2}} du$$

$$+ 6\sigma^2 \int_{-1/\sigma}^{0} u^2 e^{-\frac{u^2}{2}} du + 4\sigma \int_{-1/\sigma}^{0} u e^{-\frac{u^2}{2}} du + \int_{-1/\sigma}^{0} e^{-\frac{u^2}{2}} du$$

$$= 2\sigma^4 \sqrt{2} \int_0^{1/2\sigma^2} p^{\frac{3}{2}} e^{-p} dp + 8\sigma^3 \int_0^{1/2\sigma^2} p e^{-p} dp$$

$$+ 6\sigma^2 \sqrt{2} \int_0^{1/2\sigma^2} p^{\frac{1}{2}} e^{-p} dp + 4\sigma \int_0^{1/2\sigma^2} e^{-p} dp$$

$$+ \frac{1}{\sqrt{2}} \int_0^{1/2\sigma^2} p^{-\frac{1}{2}} e^{-p} dp \tag{27}$$

Using the results given in (17), it can be shown after a series of algebraic manipulations, that (27) is equal to

$$\sqrt{\frac{\pi}{2}}\Phi\left(\frac{1}{\sqrt{2\sigma^2}}\right)\left(1+3\sigma^2+3\sigma^4\right)$$

$$-\sigma e^{-\frac{1}{2\sigma^2}}\left(15+11\sigma^2\right)+4\sigma\left(1+2\sigma^2\right) \quad (28)$$

hence

$$\int_{-1/\sigma}^{0}(\sigma u+1)^4 e^{-\frac{u^2}{2}}du$$

$$=\sqrt{\frac{\pi}{2}}\Phi\left(\frac{1}{\sqrt{2\sigma^2}}\right)\left(1+3\sigma^2+3\sigma^4\right)$$

$$-\sigma e^{-\frac{1}{2\sigma^2}}\left(15+11\sigma^2\right)+4\sigma\left(1+2\sigma^2\right) \quad (29)$$

Furthermore applying the substitution in (19) and its corresponding results in (20a) and (20b) into

$\int_0^\infty(\sigma u+1)^4 e^{-\frac{u^2}{2}}du$, we obtain

$$\int_0^\infty(\sigma u+1)^4 e^{-\frac{u^2}{2}}du$$

$$=2\sigma^4\sqrt{2}\int_0^\infty p^{\frac{3}{2}}e^{-p}dp+8\sigma^3\int_0^\infty pe^{-p}dp$$

$$+6\sigma^2\sqrt{2}\int_0^\infty p^{\frac{1}{2}}e^{-p}dp+4\sigma\int_0^\infty e^{-p}dp+\frac{1}{\sqrt{2}}\int_0^\infty p^{-\frac{1}{2}}e^{-p}dp$$

$$=4\sigma\left(1+2\sigma^2\right)+\sqrt{\frac{\pi}{2}}\left(1+6\sigma^2+3\sigma^4\right) \quad (30)$$

Substituting the results in (29) and (30) and the value of k into (24), we have that

$$E\left(Y^2\right)=\frac{\left[8\sigma\left(1+2\sigma^2\right)+g\left(\sigma\right)-\sigma e^{-\frac{1}{2\sigma^2}}\left(15+11\sigma^2\right)\right]}{\sqrt{2\pi}\left(1-\Phi\left(-\frac{1}{\sigma}\right)\right)} \tag{31}$$

where

$$=\sqrt{\frac{\pi}{2}}\left[3\sigma^2+\left(1+3\sigma^2+3\sigma^4\right)\left(1+\Phi\left(\frac{1}{\sqrt{2\sigma^2}}\right)\right)\right]\mathrm{Var}\left(Y\right)$$

$$=\frac{\left[8\sigma\left(1+2\sigma^2\right)+g\left(\sigma\right)-\sigma e^{-\frac{1}{2\sigma^2}}\left(15+11\sigma^2\right)\right]}{\sqrt{2\pi}\left(1-\Phi\left(-\frac{1}{\sigma}\right)\right)}$$

$$-\left[\frac{2\sigma\left(1-e^{-\frac{1}{2\sigma^2}}\right)+\sqrt{\frac{\pi}{2}}\left(1+\sigma^2\right)\left(1+\Phi\left(\sqrt{\frac{1}{2\sigma^2}}\right)\right)}{\sqrt{2\pi}\left(1-\Phi\left(-\frac{1}{\sigma}\right)\right)}\right]^2 \tag{32}$$

CONDITION FOR SUCCESSFUL SQUARE TRANSFORMATION

In this section, the interval $0<\sigma\leq b\in R^+$ for which the desirable properties of the square transformed left-truncated $N\left(1,\sigma^2\right)$ distribution is approximately the same to that of the untransformed. The properties of interest are unit mean and normality and as a result we would first determine the interval $0<\sigma\leq b\in R^+$ for which the means of the transformed and the untransformed distributions of interest are both equal to unity (That is $E(X)=E(Y)=1.0$). Secondly we would determine the

Table 1: An abridged table showing the computations of E(X), E(Y), Var(X) and Var(Y).

SN	σ	E(X)	E(Y)	Var(X)	Var(Y)
1	0.0100	1.0000	1.0180	0.0001	1.0325
2	0.0110	1.0000	1.0198	0.0001	1.0358
3	0.0120	1.0000	1.0216	0.0001	1.0392
4	0.0130	1.0000	1.0234	0.0002	1.0425
5	0.0140	1.0000	1.0252	0.0002	1.0459
6	0.0150	1.0000	1.0270	0.0002	1.0492
7	0.0160	1.0000	1.0288	0.0003	1.0526
8	0.0170	1.0000	1.0306	0.0003	1.0560
9	0.0180	1.0000	1.0324	0.0003	1.0594
10	0.0190	1.0000	1.0342	0.0004	1.0628
11	0.0200	1.0000	1.0360	0.0004	1.0663
12	0.0210	1.0000	1.0378	0.0004	1.0697
13	0.0220	1.0000	1.0396	0.0005	1.0732
14	0.0230	1.0000	1.0414	0.0005	1.0767
15	0.0240	1.0000	1.0431	0.0006	1.0801
16	0.0250	1.0000	1.0449	0.0006	1.0836
17	0.0260	1.0000	1.0467	0.0007	1.0871
18	0.0270	1.0000	1.0485	0.0007	1.0907
19	0.0280	1.0000	1.0503	0.0008	1.0942
20	0.0290	1.0000	1.0521	0.0008	1.0978
21	0.0300	1.0000	1.0539	0.0009	1.1013
22	0.0310	1.0000	1.0557	0.0010	1.1049
23	0.0320	1.0000	1.0575	0.0010	1.1085
24	0.0330	1.0000	1.0593	0.0011	1.1121
25	0.0340	1.0000	1.0611	0.0012	1.1157
26	0.0350	1.0000	1.0629	0.0012	1.1193
27	0.0360	1.0000	1.0647	0.0013	1.1230

SN	σ	E(X)	E(Y)	Var(X)	Var(Y)
28	0.0370	1.0000	1.0665	0.0014	1.1266
29	0.0380	1.0000	1.0683	0.0014	1.1303
30	0.0390	1.0000	1.0701	0.0015	1.1340
31	0.0400	1.0000	1.0719	0.0016	1.1377
32	0.0410	1.0000	1.0737	0.0017	1.1414
33	0.0420	1.0000	1.0755	0.0018	1.1451
34	0.0430	1.0000	1.0773	0.0018	1.1488
35	0.0440	1.0000	1.0791	0.0019	1.1526
36	0.0450	1.0000	1.0809	0.0020	1.1564

value of $b \in R^{+}$ for which the curve shapes of the two probability distributions are bell-shaped and symmetrical about a unit mean.

For the purpose of this investigation, $E(X), Var(X)$, $E(Y)$ and $Var(Y)$ using Equations (3), (4), (22) and (32) are computed for values of $\sigma = 0.01, 0.011, 0.012, \cdots, 0.298, 0.299, 0.300$. The results of the computations are given in Table 1 (For want of space, Table 1 is an abridged table). From Table 1, the following results are true;

1) $E(X) = E(Y) \cong 1.0$ to one decimal place (dp) for the interval $\{0 < \sigma \leq 0.027; b = 0.027\}$.

2) $E(X) = E(Y) \cong 1.0$ to two decimal places for the interval $\{0 < \sigma \leq 0.028; b = 0.028\}$. In order to determine the number of decimal place(s) to use, we investigate the normality of the pdf curves of the square transformed and that of the untransformed distributions at the points b = 0.027 and 0.280. The investigation of normality at the two points is based on the previous studies of [1,12,13] whereby, normality of a pdf curve at a point b implied normality at points $0 < a \leq b \in R^{+}$.

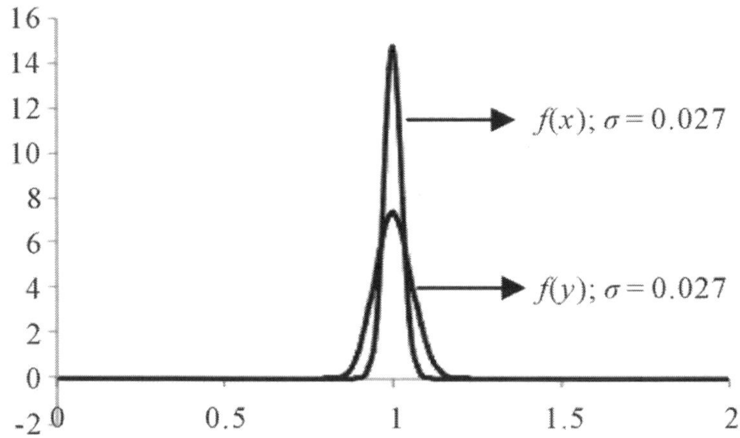

Figure 1: Curve Shapes of the Square-transformed $(f(y))$ and Un-transformed $(f(x))$ probability density function of the left-truncated $N(1,\sigma^2)$ Distribution for $\sigma = 0.027$.

Bell-shaped curves and symmetry about a unit mean would be a measure of normality. The pdf curves of the two distributions of interest for b = 0.027, 0.280 are given in Figures 1 and 2.

From the Figures, it is obvious that there is a clear departure from normality for the pdf curves for b = 0.280, therefore the acceptable interval is $0 < \sigma \leq 0.027$, since there is clear evidence of normality at b = 0.027. It is also clear from Table 1 that $Var(Y) > Var(X) \forall \sigma$. Furthermore it is also evidenced from Table 1 that $Var(Y) \cong 1.0$, $0 < \sigma < 0.027$, hence $Y \sim N(1.0,1.0), 0 < \sigma \leq 0.027$ correct to one decimal place (dp).

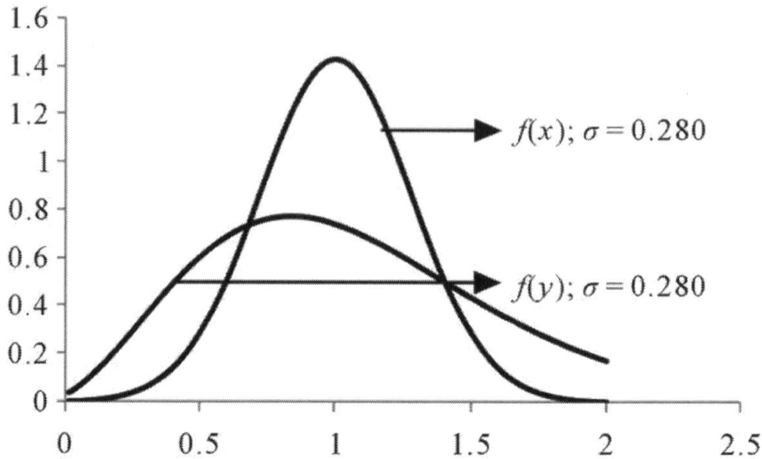

Figure 2: Curve Shapes of the Square-transformed $(f(y))$ and Un-transformed $(f(x))$ probability density function of the left-truncated $N(1,\sigma^2)$ Distribution for $\sigma = 0.280$.

SUMMARY AND CONCLUSION

In this study we have established the pdf of the square transformed left-truncated $N(1,\sigma^2)$ error component of the multiplicative time series model and the functional expressions for its mean and variance. Furthermore the mean and variance of the square transformed left-truncated $N(1,\sigma^2)$ error component and those of the untransformed component were compared for the purpose of establishing the interval for σ where the properties of the two distributions are approximately the same in terms of equality of means and normality. From the results of the study, it was established that the two distributions are normally distributed and have means $\cong 1.0$ correct to 1dp in the interval $0 < \sigma \leq 0.027$.

Based on the results of this study we therefore conclude that successful square transformation where necessary is achieved for values of σ such that $0 < \sigma \leq 0.027$. However caution has to be exercised since square transformation leads to increased error variance.

REFERENCES

1. I. S. Iwueze, "Some Implications of Truncating the N (1, s^2) Distribution to the Left at Zero," Journal of Applied Sciences, Vol. 7, No. 2, 2007, pp. 189-195. doi:10.3923/jas.2007.189.195
2. C. Chatfield, "The Analysis of Time Series: An Introduction," Chapman and Hall, CRC Press, Boca Raton, 2004.
3. D. B. Percival and A. T. Walden, "Wavelet Methods for Time Series Analysis," Cambridge University Press, Cambridge, 2000.
4. M. B. Priestley, "Spectral Analysis and Time Series Analysis," Vol. 1-2, Academic Press, London, 1981.
5. G. E. P. Box, G. M Jenkins and G. C. Reinsel, "Time Series Analysis, Forecasting and Control," 3rd Edition, Prentice-Hall, Englewood Cliffs, 1994.
6. W. W. Wei, "Time Series Analysis: Univariate and Multivariate Methods," Addison-Wesley Publishing Company, Inc., Redwood City, 1989.
7. M. G. Kendal and J. K. Ord, "Time Series," 3rd Edition, Charles Griffin, London, 1990.
8. I. S. Iwueze, E. C. Nwogu, J. Ohakwe and J. C. Ajaraogu, "Uses of the Buys-Ballot Table in Time Series Analysis," Applied Mathematics, Vol. 2, No. 5, 2011, pp. 633-645.doi:10.4236/am.2011.25084
9. M. S. Bartlett, "The Use of Transformations," Biometrica, Vol. 3, No. 1, 1947, pp. 39-52.doi:10.2307/3001536
10. G. E. P. Box and D. R. Cox, "An Analysis of Transformations," Journal of the Royal Statistical Society, Series B, Vol. 26, No. 2, 1964, pp. 211-243.
11. A. C. Akpanta and I. S. Iwueze, "On Applying the Bartlett Transformation Method to Time Series Data," Journal of Mathematical Sciences, Vol. 20, No. 3, 2009, pp. 227-243.
12. C. R. Nwosu, I. S. Iwueze and J. Ohakwe, "Distribution of the Error Term of the Multiplicative Time Series Model under Inverse Transformation," Advances and Applications in Mathematical Sciences, Vol. 7, No. 2, 2010, pp. 119-139.
13. E. L. Otuonye, I. S. Iwueze and J. Ohakwe, "The Effect of Square Root Transformation on the Error Component of the Multiplicative Time Series Model," International Journal of Statistics and Systems, Vol. 6, No. 4, 2011, pp. 461-476.
14. J. Ohakwe, O. A. Dike and A. C. Akpanta, "The Implication of Square Root Transformation on a Gamma Distributed Error Component of a Multiplicative Time Series Model," Proceedings of African Regional Conference on Sustainable Development, Vol. 6, No. 4, 2012, pp. 65-78.
15. R. V. Hogg and A. T. Craig, "Introduction to Mathematical Statistics," 6th Edition, Macmillan Publishing Co. Inc., New York, 1978.

CITATION

J. Ohakwe, O. Iwuoha and E. Otuonye, "Condition for Successful Square Transformation in Time Series Modeling," Applied Mathematics, Vol. 4 No. 4, 2013, pp. 680-687. doi: 10.4236/am.2013.44093.

Information Worth Of Minmaxent Models For Time Series

Aladdin Shamilov[1], *Cigdem Giriftinoglu*[1,2]

[1]Department of Statistics Anadolu University, Eskisehir, Turkey

[2]Department of Economics, University of Illinois, Urbana-Champaign, USA

2

ABSTRACT

In this study, by starting from Maximum entropy (MaxEnt) distribution of time series, we introduce a measure that quantifies information worth of a set of autocovariances. The information worth of autocovariences is measured in terms of entropy difference of MaxEnt distributions subject to different autocovariance sets due to the fact that the information discrepancy between two distributions is measured in terms of their entropy difference in MaxEnt modeling. However, MinMaxEnt distributions (models) are obtained on the basis of MaxEnt distributions dependent on parameters according to autocovariances for time series. This distribution is the one which has minimum entropy and maximum information out of all MaxEnt distributions for family of time series constructed by considering one or several values as parameters. Furthermore, it is shown that as the number of autocovariances increases, the entropy of approximating distribution goes on decreasing. In addition, it is proved that information worth of each model defined on the basis of MinMaxEnt modeling about stationary time series is equal to sum of all possible information increments corresponding to each model with respect to preceding model starting with first model in the sequence of models. The fulfillment of obtained results is demonstrated on an example by using a program written in Matlab.

INTRODUCTION

In many instances, the type of data available for modeling and that used for optimization is a set of observations measured over time of system variable(s) of interest [1] -[4]. A time series stated as only one realization of a stochastic process is a set of data measured through time. In many areas from engineering to economics, patterns of time series are encountered. It is difficult to find a science program not required to study with a data set in form of time series. The characteristic property of a time series is that its future behavior can not be exactly estimated. It is not uncommon in economic analysis to develop a model and perform empirical analysis by assuming that economic agents make decisions based on a set of available information [5]. In empirical analyses, however, the information is usually designated by a generic information set S. There is no attempt to quantify the amount of information in S. A quantification of the worth of such a set would not be an easy task even if one could identify all elements of S [6]. In this paper, we view the flow of information to a stochastic process from the autocovariance sets and consider measuring the amount of information when S is a set which consists of autocovariances obtained from the time series. For this reason, it is concerned with the analysis of the ordered data using the principle of maximum entropy when the information about the times series is given by autocovariances up to a lag m. According to the maximum entropy approach, given time series can be viewed as single trial from a stochastic process that is stationary up to its second-order statistics and has a zero mean. It is known that MaxEnt distribution of an observed time series is determined as a multivariate normal distribution whose dimension is equal to the number of observations [1]. By virtue of the entropy of normal distribution, entropy optimization (EO) functional is constructed as H_{max}. It can be shown that as the number of constraints generated by autocovariances increases, value of H_{max} decreases. In this investigation, firstly MaxEnt distribution for stationary time series subject to constraints generated by autocovariances set $\{r_0, r_1, \cdots, r_m\}$ is considered. It is proved that as number of lags of successive autocovariances increases, the entropy value of this distribution goes on decreasing but its information worth goes on increasing. Furthermore, by starting from MaxEnt distribution dependent on parameters, MinMaxEnt distribution which has minimum entropy and maximum information out of all MaxEnt distributions is defined. It should be noted that MinMaxEnt and MaxMaxEnt distributions as solutions of Generalized Entropy Optimization (GEO) problem firstly are defined and generally investigated in [7] -[9]. In [10] , GEO distribution dependent on parameters in time series is introduced and via this distribution an estimation method of missing value is proposed. In this study, it is shown that entropy value and information worth of MinMaxEnt distribution obtained on the bases of MaxEnt distribution dependent on parameters has the same above expressed properties at each fixed value of parameters as MaxEnt distribution. In addition, it is proved that information worth of each model defined on the basis of MinMaxEnt modeling about stationary time series is equal to the sum of all possible information increments corresponding to each model with respect to preceding model starting with first model in the sequence of models. The fulfillment of obtained results is demonstrated on an example by the use of a program written in Matlab.

INFORMATION WORTH OF AUTOCOVARIANCES SET IN MAXENT MODELING

In this section, MaxEnt distributions according to different number of autocovariances are considered and it is proved that the entropy values of these distributions constitute a monotonically decreasing sequence when the number of autocovariances increases. Moreover it is shown that the information generated by autocovariances set is expressed as sum of information worth of each autocovariance taken separately.

Theorem 1: Let $r_0, r_1, \cdots, r_m$ be autocovariances with $0, 1, \cdots, m$ lags of observed stationary time series $\{y_0, y_1, \cdots, r_n\}$, $P_{r_k}(y)$ be MaxEnt distribution subject to constraints generated by autocovariances set $\{r_k\} = \{r_0, r_1, \cdots, r_k\}$, $k = 1, \cdots, m$; $m < N$ and $H_{max}(r_k)$ be the entropy value of this distribution. Then, entropy values of mentioned MaxEnt distributions form a monotonically decreasing sequence of the following form:

$$H_{\max}(r_0) > H_{\max}(r_1) > H_{\max}(r_2) > \cdots > H_{\max}(r_m) \tag{1}$$

Proof. The Shannon entropy measure subject to constraints generated by autocovariances $r_0, r_1, \cdots, r_m$ with $0, 1, \cdots, m$ of stationary time series $\{y_0, y_1, \cdots, y_n\}$ is multivariate normal [1]. Therefore by increasing of the number of k of autocovariances vector $\{r_k\} = \{r_0, r_1, \cdots, r_k\}$, the conditions to maximize Shannon measure is increased and the domain of entropy measure becomes narrow. Consequently, entropy value of $H_{\max}(r_k)$ is strongly decreased and the inequalities (1) are satisfied. Theorem 1 is proved.

If we denote by $I(r_k)$ information worth of autocovariance r_k, due to the fact that the information discrepancy between two distributions is measured in terms of their entropy difference in MaxEnt modeling, then

$$I(r_k) = H_{\max}(r_{k-1}) - H_{\max}(r_k), \quad k = 1, \cdots, m. \tag{2}$$

Furthermore, if information worth generated by autocovariances set $\{r_1, L, r_k\}$ in the aggregate is denoted by $I(r_1, L, r_k)$, $k = 1, L, m$, then

$$I(r_1, \cdots, r_k) = H_{\max}(r_0) - H_{\max}(r_k). \tag{3}$$

Remark 1: The information $I(r_1, \cdots, r_k)$, $k = 1, \cdots, m$, generated by autocovariances set $\{r_1, \cdots, r_k\}$ is expressed as sum of information worths of each autocovariances taken separately,

$$I(r_1, \cdots, r_k) = I(r_1) + \cdots + I(r_k). \tag{4}$$

From (3) by virtue of formula (2) follows

$$\begin{aligned} I(r_1,\cdots,r_k) &= H_{\max}(r_0) - H_{\max}(r_k) \\ &= H_{\max}(r_0) - H_{\max}(r_1) + H_{\max}(r_1) - H_{\max}(r_2) + \cdots + H_{\max}(r_{k-1}) - H_{\max}(r_k) \\ &= I(r_1) + \cdots + I(r_k), \end{aligned}$$

consequently

$$I(r_1,\cdots,r_k) = I(r_1) + \cdots + I(r_k).$$

INFORMATION WORTH OF DEPENDENT ON PARAMETERS

In this section, according to different number of autocovariances MaxEnt distributions dependent on parameters are considered and it is proved that at each value of parameter, these distributions and their entropies possess the same properties as in section 2.

Theorem 2: Let $P_{r_k}(y)$ be MaxEnt distribution generated by autocovariances set $\{r_0,r_1,\cdots,r_m\}$ of given stationary time series $\{y_0,\cdots,y_{s-1},\gamma,y_{s+1},\cdots,y_n\}$ with missing value $\gamma\in[\alpha,\beta]$, where $\alpha=\min_{j\neq i}\{y_j\},\beta=\min_{j\neq i}\{y_j\}$. Since $r_0,r_1,\cdots,r_m$ depend on γ, MaxEnt distribution $P_{r_k}(y)$ is also dependent on γ. Thereafter, autocovariances set $\{r_0,r_1,\cdots,r_m\}$ will be represented as $r_0(\gamma),\cdots,r_m(\gamma)$, MaxEnt distribution as $P_{r_k(\gamma)}(y)$ and entropy of this distribution as $H_{\max}(r_k(\gamma))$. Thus, we have a family of time series dependent on γ.

Between entropy values $H_{\max}(r_k(\gamma))$ of MaxEnt distributions $P_{r_k(\gamma)}(y)$, $k=0,1.\cdots,m$ the following inequalities are fulfilled:

$$H_{\max}(r_0(\gamma)) > H_{\max}(r_1(\gamma)) > H_{\max}(r_2(\gamma)) > \cdots > H_{\max}(r_m(\gamma)) \quad (5)$$

In other words, entropy values of MaxEnt distributions dependent on γ constitute a monotonically decreasing sequence.

Proof. According to Theorem 1, entropy values $H_{max}(r_k)$ of MaxEnt distributions form a monotonically decreasing sequence of the form (1). Since $r_0(\gamma),\cdots,r_m(\gamma)$ depend on γ. Consequently, inequalities (5) are satisfied. Theorem 2 is proved.

Information worth $I(r_k(\gamma))$ of autocovariance r_k dependent on γ is determined by the following equation similarly to (2),

$$I(r_k(\gamma)) = H_{\max}(r_{k-1}(\gamma)) - H_{\max}(r_k(\gamma)),\ k = 1,\cdots,m. \tag{6}$$

Then, information worth generated by autocovariances set $\{r_1(\gamma),L,r_k(\gamma)\}$ is denoted by $I(r_1(\gamma),L,r_k(\gamma))$, $k = 1,L,m$, and

$$I(r_1(\gamma),\cdots,r_k(\gamma)) = H_{\max}(r_0(\gamma)) - H_{\max}(r_k(\gamma)),\ k = 1,\cdots,m. \tag{7}$$

Remark 2: The information $I(r_1(\gamma),L,r_k(\gamma))$, $k = 1,L,m$, generated by autocovariances set $\{r_1(\gamma),L,r_k(\gamma)\}$ is expressed as sum of information worths of each autocovariances taken separately,

$$I(r_1(\gamma),\cdots,r_k(\gamma)) = I(r_1(\gamma)) + \cdots + I(r_k(\gamma)). \tag{8}$$

From (7) by virtue of formula (6) follows

$$\begin{aligned} I(r_1(\gamma),\cdots,r_k(\gamma)) &= H_{\max}(r_0(\gamma)) - H_{\max}(r_1(\gamma)) \\ &= H_{\max}(r_0(\gamma)) - H_{\max}(r_1(\gamma)) + H_{\max}(r_1(\gamma)) - H_{\max}(r_2(\gamma)) \\ &\quad + \cdots + H_{\max}(r_{k-1}(\gamma)) - H_{\max}(r_k(\gamma)) \\ &= I(r_1(\gamma)) + \cdots + I(r_k(\gamma)) \end{aligned}$$

$$I(r_1(\gamma),\cdots,r_k(\gamma)) = I(r_1(\gamma)) + \cdots + I(r_k(\gamma)).$$

INFORMATION WORTH OF MINMAXENT MODELS DEPENDENT ON AUTOCOVARIANCES

In this section, MinMaxEnt distributions (models) are obtained on the basis of MaxEnt distributions dependent on parameters and it is shown that as the number of autocovariances k goes on increasing, the entropy of approximating distribution (model) goes on decreasing. Furthermore, it is proved that information worth of each model defined on the basis of MinMaxEnt modeling about stationary time series is equal to the sum of all possible information increments corresponding to each model with respect to preceding model starting with first model in the sequence of models.

Theorem 3: Let $P_{r_k(\gamma)}(y)$ be MaxEnt distribution generated by autocovariances set $\{r_0(\gamma), r_1(\gamma), \cdots, r_k(\gamma)\}$ of given stationary time series $\{y_0, \cdots, y_{s-1}, \gamma, y_{s+1}, \cdots, y_N\}$ with parameter $\gamma \in [\alpha, \beta]$, at position s, where $\alpha = \min_{j \neq i}\{y_j\}$, $\beta = \min_{j \neq i}\{y_j\}$ and entropy value of this distribution be $H_{max}(r_k(\gamma))$. Moreover, let γ_k, $k = 1, \cdots, m$ be the value $\gamma_k \in [\alpha, \beta]$ realizing MinMaxEnt distribution $P_{r_k(\gamma)}(y)$, in other words

$$\min_{\gamma \in [\alpha, \beta]} H_{\max}(r_k(\gamma)) = H_{\max}(r_k(\gamma_k)), \quad k = 1, \cdots, m \qquad 9)$$

Then, between entropy values of MinMaxEnt distributions the inequalities

$$H_{\max}(r_0(\gamma_0)) > H_{\max}(r_1(\gamma_1)) > H_{\max}(r_2(\gamma_2)) > \cdots > H_{\max}(r_m(\gamma_m)) \qquad (10)$$

are satisfied.

Proof. According to Theorem 2 for any $\gamma, \gamma \in [\alpha, \beta]$, the inequalities (5) hold. For this reason,

$$\min_{\gamma\in[\alpha,\beta]} H_{\max}\left(r_0\left(\gamma\right)\right) = H_{\max}\left(r_0\left(\gamma_0\right)\right) > H_{\max}\left(r_1\left(\gamma_0\right)\right) \tag{11}$$

On the other hand,

$$\min_{\gamma\in[\alpha,\beta]} H_{\max}\left(r_1\left(\gamma\right)\right) = H_{\max}\left(r_1\left(\gamma_1\right)\right) \tag{12}$$

$$\min_{\gamma\in[\alpha,\beta]} H_{\max}\left(r_1\left(\gamma_0\right)\right) > H_{\max}\left(r_1\left(\gamma_1\right)\right) \tag{13}$$

From inequality (11) by taken into account (12) and (13), the inequality

$$H_{\max}\left(r_0\left(\gamma_0\right)\right) > H_{\max}\left(r_1\left(\gamma_1\right)\right) \tag{14}$$

is got. If this process is consecutively repeated, then it is easy to get to the inequalities (10). Theorem 3 is proved.

Remark 3: By using Theorem 3, it is possible to obtain information worth of MinMaxEnt distributions with the different number of autocovariances.

By using Theorem 3, it is possible to obtain information worth of MinMaxEnt distributions with the different number of autocovariances. However, in order to simplify the description of results, we introduce the following symbols. Let Y_k, $(k=1,\cdots,m)$ be a model representing MinMaxEnt distribution $P_{r_k(\gamma_k)}(y)$ for a stationary time series $\{y_0,\cdots,y_{s-1},\gamma,y_{s+1},\cdots,y_N\}$. Moreover, let $I(k)$, $(k=1,\cdots,m)$be the information contained by model Y_k about this time series, then

$$I^{(k)} = H\left(Y_0\right) - H\left(Y_k\right),\ \ k=1,\cdots,m \tag{15}$$

and

$$I^{(k-1)} = H\left(Y_0\right) - H\left(Y_{k-1}\right),\ \ k=1,\cdots,m \tag{16}$$

From (15) and (16),

$$\begin{aligned} I^{(k)} - I^{(k-1)} &= H(Y_0) - H(Y_k) - \left(H(Y_0) - H(Y_{k-1})\right) \\ &= H(Y_{k-1}) - H(Y_k) = I_k,\ k = 1,\cdots,m \\ &= I_k \end{aligned}$$

$$I^{(k)} - I^{(k-1)} = I_k \tag{17}$$

where I_k is the information increment corresponding to each model Y_k with respect to preceding model Y_{k-1}. By virtue of the obtained results, the following theorem can be asserted.

Theorem 4: Information worth $I^{(m)}$ of model Y_m defined on the basis of MinMaxEnt modelling about stationary time series $\{y_0,\cdots,y_{s-1},\gamma,y_{s+1},\cdots,y_N\}$ is equal to sum of all possible information increments $I_k = I^{(k)} - I^{(k-1)} = H(Y_{k-1}) - H(Y_k)$, $I_k = I^{(k)} - I^{(k-1)} = H(Y_{k-1}) - H(Y_k)$ corresponding to each model with respect to preceding model Y_{k-1} starting with first model $(1 \le k \le m)$ in the sequence of models $Y_0, Y_1, \cdots, Y_m$.

Proof. By using the new notations Y_0, Y_1, L, Y_m inequalities (10) can be represented as

$$H(Y_0) > H(Y_1) > H(Y_2) > \cdots > H(Y_m) \tag{18}$$

Equation (10) shows that as the number of autocovariances k increases, the entropy of approximating distribution (model) goes on decreasing but it never goes below the entropy of probability distribution satisfying the same conditions as MinMaxEnt distribution. According to (15) and (17)

$$
\begin{aligned}
I^{(m)} &= H(Y_0) - H(Y_1) \\
&= H(Y_0) - H(Y_1) + H(Y_1) - H(Y_2) + \cdots + H(Y_{m-1}) - H(Y_m) \\
&= I_1 + I_2 + \cdots + I_m
\end{aligned}
$$

or

$$
I^{(m)} = I_1 + I_2 + \ldots + I_m \tag{19}
$$

According to (18) in (19), $I_k > 0, k = 1.\cdots, m$. Theorem 4 is proved.

APPLICATIONS

The developed MinMaxEnt models $Y_0, Y_1, \cdots, Y_m$ can be applied to estimate the missing value in time series. According to Theorem 4, information worth generated by Y_m is greater than information worth generated by Y_{m-1}. Consequently, γ_m generating the model Y_m is the better estimation than γ_{m-1} generating the model Y_{m-1} in the sense of information worth. On An example it is shown that mentioned estimated value is the best also in the sense of mean square error (MSE). To realize required operations, a program in MATLAB is written. For this purpose, we have considered data set generated from autoregressive process $AR(4)$ as follows:

$$
X_t = 1.90X_{t-1} - 2.01X_{t-2} + 1.84X_{t-3}6 - 0.80X_{t-4} + \varepsilon_t,\ \varepsilon_t \sim N(0, 0.5) \tag{20}
$$

and the data set is given in Table 1. By using the data in Table 1, estimations based on MinMaxEnt models are obtained for missing values in each position via constraints generated by r_0, r_1, r_2 autocovariances and r_0, r_1, r_2, r_3 autocovariances.FromTable1itisseenthat,MinMaxEntestimations γ_{3t} determined by the set consisting of r_0, r_1, r_2, r_3 autocovariances are

Table 1: The data generated from AR(4) and its estimations with different autocovariance sets.

t	X(t)	γ_{2t}	γ_{3t}	t	X(t)	γ_{2t}	γ_{3t}
1	-7.6164	-3.7049	-4.5863	26	-2.5809	-0.4340	-12861
2	-7.9251	-5.9152	-8.2637	27	-1.8546	0.0249	-1.8094
3	-23466	-3.1335	-1.5912	28	4.7113	2.7239	4.5856
4	-1.0728	-2.6884	-1.1953	29	5.2406	2.9464	5.1481
5	-63050	-4.4728	-6.1961	30	-1.0943	0.4262	-0.8107
6	-7.7206	-49192	-7.7193	31	-2.4052	-0.1378	-23785
7	-2.2376	-2.6242	-2.2308	32	4.0709	3.0661	43309
8	033865	-1.6810	-0.0090	33	7.9505	4.9433	7.7333
9	-4.5611	-32248	-4.3121	34	3.5777	3.7644	35149
10	-7.3435	-4.7417	-7.6510	35	02252	3.1348	0.8623
11	-33723	-2.9111	-32169	36	-2.4052	-0.1378	-23785
12	0.13548	-1.8088	-0.0447	37	4.0709	3.0661	43309
13	-3.7786	-3.4259	-3.7174	38	7.9505	4.9433	7.7333
14	-8.2637	-5.3028	-8.1113	39	3.5777	3.7644	3.5249
15	-5.2458	-4.0749	-4.8305	40	0.8252	3.1348	0.8623
16	-02230	-2.2286	-0.1069	41	11292	8.1259	11.159
17	-2.1272	-2.2977	-1.8858	42	7.5889	6.6536	7.3807

t	X(t)	γ_{2t}	γ_{3t}	t	X(t)	γ_{2t}	γ_{3t}
18	-5.4257	-2.6645	-52509	43	3.3139	5.2987	3.5224
19	-1.0920	-0.1997	-13106	44	6.5842	6.0319	6.2192
20	5.5526	3.1233	5.2295	45	10.412	75539	10.267
21	4.5110	3.1064	3.9525	46	7.2051	6.1065	7.1059
22	-08572	1.2503	-09899	47	2.0869	3.6081	2.1044
23	0.0716	1.1921	-0.2413	48	3.1468	3.5739	3.1619
24	4.7447	2.7483	4.8959	49	7.1153	5.0748	7.1544
25	3.4163	1.6973	3.3217	50	5.9239	4.4184	4.9659

Table 2: Entropy and information worth of different autocovariance sets.

$H_{max}(r_0)$	$H_{max}(r_0, r_1)$	$H_{max}(r_0, r_1, r_2)$	$H_{max}(r_0, r_1, r_2, r_3)$	$I(r_1)$	$I(r_2)$	$I(r_3)$	$I(r_1, r_2, r_3)$
171.94	155.74	153.80	117.37	16.20	1.94	36.43	54.57

better than MinMaxEnt estimations γ_{2t} determined by the set consisting of r_0, r_1, r_2 autocovariances in each position. Moreover, $(MSE)_{r_0, r_1, r_2, r_3}$ calculated by MinMaxEnt estimations with autocovariances is 0.2564 and it is lower than $(MSE)_{r_0, r_1, r_2} = 3.6605$ calculated by MinMax-Ent estimations with r_0, r_1, r_2 autocovariances and $(MSE)_{r_0, r_1} = 8.0426$ calculated by MinMaxEnt estimations with r_0, r_1 autocovariances.

Furthermore, in Table 2 the entropy and information worth of different autocovariance sets are given. These quantities calculated from the data set verify Theorem 4. It can be seen that as the number of constraints which is generated by autocovariances increases, the value of H_{max} decreases.

CONCLUSIONS

In this study, the following results are established.

- MaxEnt distributions according to different number of autocovariances are considered and it is proved that the entropy values of these distributions constitute a monotonically decreasing sequence when the number of autocovariances increases. Moreover it is shown that the information generated by autocovariances set is expressed as sum of information worth of each autocovariance taken separately.
- According to different number of autocovariances, MaxEnt distributions dependent on parameters are considered and it is proved that at each value of parameter these distributions and their entropies possess the same properties as the MaxEnt distributions.
- MinMaxEnt distributions (models) are obtained on the basis of MaxEnt distributions dependent on parameters and it is shown that as the number of autocovariances k goes on increasing, the entropy of approximating distribution (model) goes on decreasing. Furthermore, it is proved that information worth of each model defined

on the basis of MinMaxEnt modeling about stationary time series is equal to the sum of all possible information increments corresponding to each model with respect to preceding model starting with first model in the sequence of models.

- Information worth of autocovariances in time series and values generating MinMaxEnt distributions can be applied in solving many problems. One of the mentioned problems is the problem of estimation of missing value in time series. It is proved that the value generating MinMaxEnt distribution independence on position represents the best estimation of the missing value in the sense of information worth.
- The fulfillment of the obtained results is demonstrated on an example by using a program written in Matlab.

ACKNOWLEDGMENTS

We thank the Editor and the referee for their comments. This support is greatly appreciated.

REFERENCES

1. Kapur, J.N. and Kesavan, H.K. (1992) Entropy Optimization Principles with Applications. Academic Press, New York.
2. Wei, W.S. (2006) Time Series Analysis, Univariate and Multivariate Methods. Pearson, United States.
3. Box, G.E.P. and Jenkins, G. (1976) Time Series Analysis: Forecasting and Control. Holden-Day, United States.
4. Little, R. and Rubin, D. (1987) Statistical Analysis with Missing Data. Wiley, New York.
5. Pourahmadi, M. and Soofi, E. (1998) Prediction Variance and Information Worth of Observations in Time Series. Journal of Time Series Analysis, 21, 413-434. http://dx.doi.org/10.1111/1467-9892.00191
6. Pourahmadi, M. (1989) Estimation and Interpolation of Missing Values of a Stationary Time Series. Journal of Time Series Analysis, 10, 149-169. http://dx.doi.org/10.1111/j.1467-9892.1989.tb00021.x
7. Shamilov, A. (2006) A Development of Entropy Optimization Methods. WSEAS Transaction on Mathematics, 5, 568- 575.

8. Shamilov, A. (2007) Generalized Entropy Optimization Problems and the Existence of Their Solutions. Physica A: Statistical Mechanics and its Applications, 382, 465-472.http://dx.doi.org/10.1016/j.physa.2007.04.014
9. Shamilov, A. (2010) Generalized Entropy Optimization Problems with Finite Moment Functions Sets. Journal of Statistics and Management Systems, 13, 595-603,.http://dx.doi.org/10.1080/09720510.2010.10701489
10. Shamilov, A. and Giriftinoglu, C. (2010) Generalized Entropy Optimization Distributions Dependent on Parameter in Time Series. WSEAS Transactions on Information Science and Applications, 1, 102-111.

CITATION

Shamilov, A. and Giriftinoglu, C. (2015) Information Worth of MinMaxEnt Models for Time Series. Applied Mathematics, **6**, 221-227. doi: 10.4236/am.2015.62021.

On the Quantum Zeno Effect and Time Series Related to Quantum Measurements

Karl-Heinz Fichtner*[1] *and
***Kei Inoue*[2]**

[1]School of Mathematics and Computer Science, Institute of Applied Mathematics, Friedrich-Schiller-Universität Jena, Jena, Germany

[2]Department of Electrical Engineering, Faculty of Engineering, Tokyo University of Science, Yamaguchi, Sanyoonoda, Japan

ABSTRACT

Our main aim is to prove a more general version of the quantum Zeno effect. Then we discuss some examples of the quantum Zeno effect. Furthermore, we discuss a possibility that based on the quantum Zeno effect and certain experiments one could check whether, from the statistical point of view, a concrete system behaves like a quantum system. The more general version of quantum Zeno effect can be helpful to prove that the brain acts like in a quantum system. The proof of our main result is based on certain formulas describing probability distributions of time series related to quantum measurements.

INTRODUCTION

Zeno of Elea (ca. 490 BC-ca. 430 BC) was one of the oldest Greek philosophers. Aristotle called him the inventor of the dialectic. He is best known for his paradoxes. The most famous are the so called "arguments against motion" described by Aristotle in his Physics. Let us mention Zeno's arrow paradox which states that, since an arrow in flight is not seen to move during any single instant, it cannot possibly be moving at all. What does it mean? Let $X(0)$ be the position of arrowhead at time $t_0 = 0$. Then the arrow will move. $X(t)$ Denotes the position of the arrowhead at time t. Now, let $t_1 < t_2 < ... < t_n$ be a sequence of times. Then the observed positions of the arrowhead are represented by the sequence $X(t_0),...X(t_n)$. If the distance of times $t_{k+1} - t_k \approx 0$ is very small then the distance of the corresponding positions $X(t_{k+1}) - X(t_k) \approx 0$ is very small. According to Zeno that implies that the arrow cannot move under continuous observation. Of course we know that this is non-sense because in classical mechanics observations are of no effect on the state of the system—not so in quantum mechanics where observation, i.e., measurement, causes a change of the state. That is the basic idea of the quantum Zeno effect.

The quantum Zeno effect is a name coined by George Sudarshan and Baidyanath Mirsa of the University of Texas in 1977 in their analysis of the situation in which an unstable particle, if observed continuously, will never decay [1]. One can "freeze" the evolution of the system by measuring it frequently enough in its (known) initial state. The meaning of the term has since expanded, leading to a more technical definition in which the time evolution can be suppressed not only by measurement [2,3]. That does not mean that there exists a complete theory of the quantum Zeno effect. Even there is no precise meaning concerning the notion of "suppression of the time evolution".

An earlier theoretical exploration of this effect of measurement was published in 1974 by Degasperis et al. [4] and A. Turing described it in 1954 [5]: It is easy to show (but there was no rigorous proof!) using standard theory that if a system starts in an eigenstate of some observable, and measurements are made of that observable N times a second, then even if the state is not a stationary one, the probability that system will be in the same state after, say, one second, tends to one as N tends to infinity; that is, that continual observations will prevent motion.

The idea is contained in the early work by J. von Neumann, sometimes called the reduction postulate [6]. According to the reduction postulate, each measurement causes the wave function to "collapse" to a pure eigenstate of the measurement basis.

Usually the model of Turing is far from the reality, because in reality the state of the quantum system is usually unknown, and it is a mixed state. Furthermore that kind of measurements is not realistic, i.e., real measuring instruments cannot perform such kind of measurements. Nevertheless a general theory of the quantum Zeno effect should contain the Turing model as a special case.

Following A. Turing the reduction of the states seems to be the background of the quantum Zeno effect. Now our work on the quantum Zeno effect was motivated by the quantum model of certain brain activities developed in [7-15]. For that reason let us briefly mention what kind of reduction of states was considered in certain quantum models of brain activities.

In Wikipedia, the free encyclopaedia on the Zeno effect is written the following:

Significance to Cognitive Science

The quantum Zeno effect is becoming a central concept in exploration of controversial theories of quantum mind consciousness within the discipline of cognitive science. In his book Mindful Universe (2007) (see also [16]), Henry Stapp claims that the mind holds the brain in a superposition of states using the quantum Zeno effect.

Now, let us consider the process of recognition of signals in the brain. Then the basic idea of Stapp may be described as follows: If there will be a signal arising from the senses, the process of recognition starts. That process represents a rapid sequence of "trials" checking whether the signal from the senses and the signal created by the brain partially coincide. He identifies these trials with measurements performed by a mysterious "observer" in the brain. That does not seem to be realistic.

Related to the model of recognition discussed in [7-15], these trials are not represented by a process of measurements. Nevertheless the results of the trials are represented by projections like in the case of measurements.

That R. Penrose [17,18] explains as follows: The conventional quantum theory view is that the quantum state reduces by measurement or observation (subjective reduction). Now, a number of physicists have argued in support of special models in which the rules of standard quantum mechanics are modified by the inclusion of some additional procedure according to which the reduction of the state becomes an objectively real process (objective reduction)—the system abruptly self-collapses. Consciousness, it is argued, requires non-computability. The only readily available apparent source of non-computability is self-collapse. The self-collapse, irreversible in time, creates an instantaneous "now" event. Sequences of such events create a flow of time, and consciousness.

That means recognition of signals is a process of self-collapses. That coincides with the basic idea of the quantum model of recognition of signals in the brain in [7-15]. Furthermore, like sequences of measurements that process suppresses a certain unitary time evolution in the brain. A lot of experiments show that there is a suppression of a certain process in the brain caused by the recognition of signals or some other kind of activities of the brain (cf. [19]).

Using series of measurements at different times one tries to get information on the statistical behaviour of certain random processes. That series of measurements give a so called time series.

Time Series—Classical Probability Theory

In classical stochastics one identifies the time evolution of a random system with a stochastic process $(\xi_t)_{t\geq 0}$ taking on values in a certain space G. The random variable ξ_t represents the system at time t. A measurement at a certain time $t \geq 0$ one identifies with a certain (measurable) real function f on G, and the random number $f(\xi_t)$ represents the output of the measurement.

Now, let us consider a sequence of time $t_0 = 0 \leq t_1 < t_2 < ... < t_n$. At time t_k one performs a measurement related to a function f_k on G. Then the sequence $\eta_k := f_k(\xi_{tk})(k = 1,...,n)$ represents a so called time series.

If we would know the distribution of the underlying stochastic process, then we can calculate the distribution of the time series.

Sometimes specialists use the notion "time series" only in the special case:

$\tau = t_k - t_{k-1;} f_k = f\ (k - 1,...,n)$.

That means one performs always the same measurements at equidistant times.

In order to obtain information concerning the state of the brain and its time evolution one has to perform several EEG measurements at different times. That gives a time series. In practice specialist's use certain statistical methods evaluating that time series, i.e., they analyze the outcomes of the sequence of measurements. To make these statistical investigations effective and precise from mathematical point of view one should know the type of the distribution of the stochastic process on the space of point configurations representing the time evolution of the configuration of excited neurons in the brain. Up to now that information is lacking (cf. [20]).

The Quantum Case of Time Series

Let $(V_t)_{t \geq 0}$ be a semigroup of unitary operators on a separable Hilbert space H representing the time evolution of a quantum system. ρ denotes the state of the quantum system at time $t_0 = 0$. We consider again a sequence of time $t_0 \le t_1 < ... < t_n$ and perform measurements at time $t_k\,(k=1,...,n)$ related to selfadjoint operators B_k given by their unique representation $B_k = \sum_{j=1}^{m_k} z_{k,j} Pr_{H_{k,j}}$, i.e., $(H_{k,j})_{j=1}^{mk}$ represents an orthogonal decomposition of H and $z_{k,j} \ne z_{k,s}$ if $j \ne s$. Because each realistic measuring instrument has a finite scale we consider only selfadjoint operators being of this type.

Formally we can identify the random outputs of these measurements with random numbers $\eta_1,...,\eta_n$ representing the time series in the considered quantum case.

Unfortunately, in the quantum case there does not exist a stochastic process $(\xi_t)_{t\ge 0}$ such that we can always represent the distribution of the time series as follows:

$$P_{\eta 1,...,\eta n} = P_{f1(\xi_{t1}),...fn(\xi_{t_n})}$$

For that reason one has to develop a useful concept in order to describe the distribution of such time series related to quantum measurements. That will be done in Section 2 of this paper.

On the Quantum Zeno Effect

Starting with the notations used above we will specialize

$$B_k = B := \sum_{j=1}^{m} z_j \, Pr_{H_j};\ t_k := \frac{k-1}{n} T\ (k=1,...,n).$$

That means we consider repeated measurements in the time interval $[0,T]$ according to B starting at time $t_1 = t_0 = 0$ with equidistant times. Again ρ denotes the state of the quantum system at time 0, and the random numbers $\eta_1,...,\eta_n$ represent the outputs of the measurements.

In Section 3 of this paper we will prove the following theorem.

Theorem It holds

$$P(\eta_k = \eta_1; k = 2,\dots,n) \xrightarrow[n\to+\infty]{} 1.$$

Immediately from the theorem we obtain the following corollary.

Corollary It holds

$$P(\eta_k = z_j; k = 1,2,\dots,n) \xrightarrow[n\to\infty]{} P(\eta_1 = z_j) = tr\left(\rho Pr_{H_j}\right),\ j = 1,\dots,m$$

That does not mean that the state will not be changed in time. This is related to the fact that using only one kind of measurement one cannot determine the state. One can consider a simple example illustrating that. Nevertheless in some special cases one has a total suppression of the time evolution in this sense that the state is not changing.

Example—A. Turing

Remember the statement of A. Turing: If a system starts in an eigenstate of some observable, and measurements are made of that observable N times a second, then the probability that the system will be in the same state after, say, one second, tends to one as N tends to infinity.

That we obtain specializing as follows: Let $\rho = Pr_\Psi$ be a state related to a normalized $\Psi \in H$, and let the measurements be related to the selfadjoint operator $B = bPr_\Psi + \hat{B}$, where $\hat{B}$ is a selfadjoint operator such that $\hat{B}$ and Pr_Ψ are orthogonal, and the eigenvalues of $\hat{B}$ are different from b. Then $\rho = Pr_\Psi$ represents an eigenstate of the operator B related to the eigenvalue b.

Now, one has to perform measurements according to B at the times

$$t_k = \frac{k-1}{N} \; (k = 1,\ldots,N+1)$$

Then we have $t_0 = 0, t_{N+1} = T$. Again η_k denotes the random output of the measurement at time $t_k (k = 1,\ldots N+1)$.

Then using the corollary mentioned above we obtain

$$P(\eta_k = b; k = 1,\ldots,N+1) \xrightarrow[n\to+\infty]{} P(\eta_1 = b)$$

Because ρ is an eigenstate of B related to the eigenvalue b we have

$$P(\eta_1 = b) = 1.$$

Thus we can conclude

$$P(\eta_k = b; k = 1,\ldots,N+1) \xrightarrow[n\to+\infty]{} 1.$$

That implies

$$P(\eta_{N+1} = b) \xrightarrow[n\to+\infty]{} 1.$$

Because of $t_{N+1} = T$ the random value η_{n+1} would represent the output of a measurement at time T and the event "$\eta_{n+1} = b$"means that at time T the system would be in the initial state Pr_Ψ. Thus we can conclude that the probability that at time T the system will be in the same state as at the beginning tends to one as N tends to infinity.

Example—Arrow Paradox

In quantum mechanics one uses the Hilbert space $H = L^2(R^3)$ in order to describe a quantum particle located in the space R^3. Now, again let $\rho = Pr_\Psi$ be a pure state related to a normalized wave function $\Psi \in H$.

We interpret ρ as the state of the arrowhead at time $t_0 = 0$, and we assume that the wave function Ψ is concentrated on a "very small region" D, thus the position of the arrowhead at time 0 is "almost fixed".

Now, let H_D be the space of the wave functions concentrated on D. We specialize $B := Pr_{H_D}$. Then the corresponding measurement means: "checking whether the arrowhead will be inside of the region D or not". Consequently the event "$\eta_k = 1$" means that "at time t_k the arrowhead is located inside of the region D".

Because of $P(\eta_1 = 1) = 1$ the corollary mentioned above may be rewritten as follows

$$P(\text{"at time } t_k \text{ the arrowhead is located in } D (k = 1,\ldots,n)\text{"}) \xrightarrow[n \to +\infty]{} 1$$

i.e., the arrow will not move under "continual observation".

Can We Use the Quantum Zeno Effect in Order to "Prove" That the Brain Acts Like a Quantum System?

More and more specialists in life science are convinced that one should use quantum models in order to describe some special bio-systems, others are against quantum models. Now we want to propose a certain experiment that may be helpful to decide whether a concrete system behaves like a quantum system or not.

The basic idea is as follows: First perform a sequence of measurements with very high frequency, i.e., the length n of the time series $\eta_1,\ldots,\eta_n$ obtained within, say, one second is very large. Then produce a time series $\hat{\eta}_1,\ldots,\hat{\eta}_m$ where m is very small (e.g. $m = 2$). Now, compare the correlations $cor(\eta_1, \eta_m)$ and $cor(\hat{\eta}_1, \hat{\eta}_m)$. In the idealistic quantum case it would be $\text{cor}(\eta_1, \eta_m) \approx 1$ because of the quantum Zeno effect. But that is not realistic because of certain classical noise caused by the measuring procedure. Nevertheless, in the quantum case $|\text{cor}(\hat{\eta}_1, \hat{\eta}_m)|$ should be significant lower than $|\text{cor}(\eta_1, \eta_m)|\_$ which would contradict to the classical case.

Finally let us mention that some parts of this paper are based on [21].

DISTRIBUTION OF A TIME SERIES RELATED TO QUANTUM MEASUREMENTS

Let $(V_t)_{t \geq 0}$ be a semigroup of unitary operators on a separable Hilbert space H representing the time evolution of a quantum system. ρ Denotes the state of the quantum system at time $t_0 = 0$.

We consider a sequence of times $t_0 \leq t_1 < \ldots < t_n$ and perform measurements at time t_k $(k = 1, \ldots, n)$ related to selfadjoint operators B_k given by their unique representation

$$B_k := \sum_{j=2}^{m_k} Z_{k,j} \, Pr_{H_{k,j}}, \, k = 1, \ldots, n$$

i.e $(H_{k,j})_{j=1}^{m_k}$ represents an orthogonal decomposition of H and $z_{k,j} \neq z_{k,s}$ if $j \neq s (k = 1, \ldots, n)$. Formally the random outputs of these measurements we identify with random numbers $\eta_1, \ldots, \eta_n$.

Theorem 1 let ρ be a pure state related to $x \in H, \|x\| = 1$. Further we put $\tau_k := t_k - t_{k-1} (k - 1, \ldots, n)$. Then we get the probabilities

$$P\left(\eta_1 = z_{1,j_1},\dots,\eta_n = z_{n,j_n}\right)$$
$$= \left\|\left(\Pr_{H_n \cdot j_n} V_{\tau_n}\right)\dots\left(\Pr_{H_n \cdot j_n} V_{\tau_1}\right)x\right\|^2$$
$$j_1 = 1,\dots,m_1,\dots,j_n = 1,\dots,m_n.$$

Proof 1 We put

$$p_k\left(z_{1,j_1},\dots,z_{k,j_k}\right)$$
$$:= \left\|\left(\Pr_{H_{k,j_k}} V_{\tau_k}\right)\cdots\left(\Pr_{H_{1,j_1}} V_{\tau_1}\right)x\right\|^2,$$
$$j_1 = 1,\cdots,m_1,\cdots,j_k = 1,\cdots,m_k; k = 1,\cdots,n. \tag{2.1}$$

Then we get

$$P\left(\eta_1 = z_{1,j_1}\right) = p_1\left(z_{1,j_1}\right), j_1 = 1,\dots,m_1. \tag{2.2}$$

Now let us assume that for some k it holds

$$P\left(\eta_1 = z_{1,j_1},\cdots,\eta_k = z_{k,j_k}\right)$$
$$= p_k\left(z_{1,j_1},\cdots,\eta_k = z_{k,j_k}\right),$$
$$j_1 = 1,\cdots,m_1,\cdots,j_k = 1,\cdots,m_k. \tag{2.3}$$

Using J. von Neumann's rule we can conclude that under condition "$\eta_1 = z_{1,j1},\dots\eta_k = z_{k,jk}$" (i.e., knowing the exact results of the k measurements) immediately after the k-th measurements the quantum system is in the pure state

$$\frac{\left(\Pr_{H_k \cdot j_k} V_{\tau_k}\right)\dots\left(\Pr_{H_1 \cdot j_1} V_{\tau_1}\right)x}{\left\|\left(\Pr_{H_k \cdot j_n} V_{\tau_n}\right)\dots\left(\Pr_{H_1 \cdot j_1} V_{\tau_1}\right)x\right\|}.$$

Then at time t_{k+1} the system is in the state

$$\frac{V_{\tau_{k+1}}\left(Pr_{H_k \cdot j_k} V_{\tau_k}\right)\dots\left(Pr_{H_1 \cdot j_1} V_{\tau_1}\right)x}{\left\|\left(Pr_{H_k \cdot j_n} V_{\tau_n}\right)\dots\left(Pr_{H_1 \cdot j_1} V_{\tau_1}\right)x\right\|}.$$

Now performing the measurement according to B_{k+1} we get the value $z_{k+1,jk+1}$ with (conditional) probability

$$P\left(\eta_{k+1} = z_{k+1,j_{k+1}} \middle| \eta_1 = z_{1,j_1}, \cdots, \eta_k = z_{k,j_k}\right)$$
$$= \frac{\left\|\left(\Pr_{H_{k+1,j_{k+1}}} V_{\tau_{k+1}}\right)\dots\left(\Pr_{H_{1,j_1}} V_{\tau_1}\right)x\right\|^2}{\left\|\left(\Pr_{H_{k,j_k}} V_{\tau_k}\right)\dots\left(\Pr_{H_{1,j_1}} V_{\tau_1}\right)x\right\|^2}. \quad (2.4)$$

Using (2.1), (2.3) and (2.4) we obtain

$$P\left(\eta_1 = z_{1,j_1}, \cdots, \eta_{k+1,j_{k+1}}\right) = p_{k+1}\left(z_{1,j_1}, \cdots, z_{k+1,j_{k+1}}\right),$$
$$j_1 = 1, \cdots, m_1, \cdots, j_{k+1} = 1, \cdots, m_{k+1}.$$

That proves Theorem 1.

Now let $f_1, \dots f_n$ be measurable functions. We will consider the expectation

$$E\prod_{r=1}^{n} f_r\left(\eta_r\right) = \sum_{j_1,\cdots,j_n} p_n\left(z_{1,j_1}, \cdots, z_{n,j_n}\right)\prod_{r=1}^{n} f_r\left(z_{r,j_r}\right). \quad (2.5)$$

Example 1 Let $D_1,..D_n$ be measurable subsets of R. We consider the indicator functions $f_k = X_{D_k}(k-1,...,n)$. Then we have

$$E\prod_{r=1}^{n} f_r(\eta_r) = P(\eta_1 \in D_1,...\eta_n \in D_n).$$

Example 2 we consider the case $f_k(y) = y^s, (r \neq k) f_r \equiv 1$. Then we have

$$E\prod_{r=1}^{n} f_r(\eta_r) = E\eta_k^s.$$

Example 3 we consider the case $f_k(y) = y^s, (r \neq k) = y, f_r \equiv 1(j \neq r \neq k)$. Then we have

$$E\prod_{r=1}^{n} f_r(\eta_r) = E\eta_k \eta_j.$$

Using Example 2 and Example 3 we can calculate correlations.

Example 4 we consider the case

$$f_r(y) = e^{is_r y}(r = 1,...,n).$$

Then

$$\gamma(S_1,...,S_n) = E\prod_{r=1}^{n} f_r(\eta_r)$$

Represents the characteristic function of the vector $[\eta_1,...,\eta_n]$. Observe that these characteristic functions represent the complete information concerning the distribution of time series

Now in quantum mechanics expectations are of the type $tr(\rho A)$ where ρ represents the state of the system and A is a certain (selfadjoint)

operator. For that reason we will construct an operator A related to $B_{1,}t_1, f_1, \ldots B_n, t_n, f_n$ such that we obtain

$$tr(\rho A) = E\prod_{r=1}^{n} f_r(\eta_r). \tag{2.6}$$

For that purpose we define the following linear mappings from the set $\hat{B}(H)$ of bounded operators on H into this set putting

$$K_t(B) := V_t^* B V_{t,} \quad B \in \hat{B}(H), t \geq 0, \tag{2.7}$$

$$W_{B_k, f_k}(B) := \sum_{j=1}^{m_k} f_k(z_{k,j}) \Pr_{H_{k,j}} B \Pr_{H_{k,j}},$$

$$B \in \hat{B}(H); k = 1, \cdots, n. \tag{2.8}$$

Then we put

$$A := A_{B_1, t_1, f_1, \cdots, B_n, t_n, f_n}$$

$$:= K_{\tau_1} \circ W_{B_1, f_1} \circ \cdots \circ K_{\tau_{n-1}} \circ W_{B_{n-1}, f_{n-1}}\left(K_{\tau_n}\left(f_n(B_n)\right)\right). \tag{2.9}$$

If the operator B is selfadjoint then $K_t(B)$ is selfadjoint. Further one easily checks that $W_{Bk,fk}(B)$ is selfadjoint if B is selfadjoint and f_k is a real function. For that reason we can conclude that the operator A is selfadjoint if $f_k (k = 1, \ldots n)$ are real functions.

Immediately from the definition of $A_{B_1, t_1, f_1, \ldots B_n, t_n, f_n}$ we get

$$A_{B_1, t_1, f_1, \cdots, B_n, t_n, f_n} = K_{t_1}\left(A_{B_1, 0, f_1, B_2, t_2 - t_1, f_2, \cdots, B_n, t_n - t_1, f_n}\right). \tag{2.10}$$

Finally we obtain the following theorem.

Theorem 2 Let the operator A be given by (2.9). Then (2.6) holds.

Proof 2 we can restrict ourselves to the case of a pure state related to $x \in H \|x\| = 1$. Then we have to prove that it holds

$$\langle x | Ax \rangle = E \prod_{r=1}^{n} f_r(\eta_r) \tag{2.11}$$

Using Theorem 1 and (2.5) we get

$$E \prod_{r=1}^{n} f_r(\eta_r) = \sum_{j_1,\cdots,j_n} \left(\prod_{r=1}^{n} f_r\left(z_{r,j_r}\right) \right) \cdot \left\| \left(\Pr_{H_{n,j_n}} V_{\tau_n} \right) \cdots \left(\Pr_{H_{1,j_1}} V_{\tau_1} \right) x \right\|^2 . \tag{2.12}$$

Further we have

$$\begin{aligned} &\left\| \left(\Pr_{H_{n,j_n}} V_{\tau_n} \right) \cdots \left(\Pr_{H_{1,j_1}} V_{\tau_1} \right) x \right\|^2 \\ &= \left\langle x \middle| \left(\Pr_{H_{1,j_1}} V_{\tau_1} \right)^* \cdots \left(\Pr_{H_{n,j_n}} V_{\tau_n} \right)^* \right. \\ &\qquad \left. \left(\Pr_{H_{n,j_n}} V_{\tau_n} \right) \cdots \left(\Pr_{H_{1,j_1}} V_{\tau_1} \right) x \right\rangle \\ &= \left\langle x \middle| V_{\tau_1}^* \Pr_{H_{1,j_1}} \cdots V_{\tau_{n-1}}^* \Pr_{H_{n-1,j_{n-1}}} V_{\tau_n}^* \Pr_{H_{n,j_n}} \right. \\ &\qquad \left. V_{\tau_n} \Pr_{H_{n-1,j_{n-1}}} V_{\tau_{n-1}} \cdots \Pr_{H_{1,j_1}} V_{\tau_1} x \right\rangle . \end{aligned} \tag{2.13}$$

Finally using (2.13), (2.12), (2.9), (2.8) and (2.7) we obtain (2.11). That proves theorem 2.

Remark 1 the time evolution of a quantum mechanical system usually is characterized by a semi group $(V_t)_{t\geq 0}$ of unitary operators. Then the state ρ_r of the system at time t is characterized by

$$\operatorname{tr}(\rho_t B) = \operatorname{tr}(\rho K_t(B)),\ B \in \hat{B}(H), t \geq 0. \tag{2.14}$$

If in addition to that time evolution there are measurements like in the case we have considered this "perturbed" evolution in general no longer may be described by a semi group of unitary operators. Though the ansatz (2.14) still makes sense. One has to replace the semi group $(K_t)_{t\geq 0}$ by a family of completely positive linear maps $\hat{K}_t$ on $\hat{B}(H)$. In the case considered above these maps are defined as follows:

$$\hat{K}_t(B) := \begin{cases} K_t(B), & 0 \leq t < t_1, \\ K_{\tau_1} \circ W_{B_1,1} \circ \cdots \circ K_{\tau_m} \circ W_{B_m,1}\left(K_{t-\tau_m}(B)\right), & t_m \leq t < t_{m+1}, \quad B \in \hat{B}(H) \\ K_{\tau_1} \circ W_{B_1,1} \circ \cdots \circ K_{\tau_n} \circ W_{B_n,1}\left(K_{t-\tau_n}(B)\right), & t_n \leq t, \end{cases} \tag{2.15}$$

Further let us mention that one can generalize our considerations replacing from the beginning the semi group $(K_t)_{t\geq 0}$ by an arbitrary semi group of identity preserving completely positive linear maps on $\hat{B}(H)$.

Remark 2 let us consider the trivial case of a time evolution

$$K_t(B) = B,\ B \in \hat{B}(H). \tag{2.16}$$

Further we consider the case of only two measurements, i.e., $n = 2$. Then we obtain the probabilities

$$P_{B_1,B_2}\left(z_{1,j_1}, z_{2,j_2}\right) = \left\| Pr_{H_2,j_1} Pr_{H_1,j_2} x \right\|^2, j_1 = 1,\ldots,m_1; j_2 = 1,\ldots,m_2. \tag{2.17}$$

The procedure may be interpreted as follows: first perform a measurement according to B_1 and then perform a measurement according to B_2.

Now we change that order, i.e., first perform a measurement according to B_2 and then perform a measurement according to B_1. Then we obtain the probabilities

$$P_{B_2,B_1}\left(z_{1,j_1},z_{2,j_2}\right)=\left\|Pr_{H_{2,j_1}}Pr_{H_{1,j_2}}x\right\|^2, j_1=1,\ldots,m_1; j_2=1,\ldots,m_2. \tag{2.18}$$

If B_1 and B_2 commute then we get like in the classical case

$$P_{B_1,B_2}\left(z_{1,j_1},z_{2,j_2}\right)=P_{B_2,B_1}\left(z_{1,j_1},z_{2,j_2}\right) j_1=1,\ldots,m_1; j_2=1,\ldots,m_2 . \tag{2.19}$$

In that case specialists say (cf. [22, 23]) that "there exist joint probabilities". If B_1 and B_2 do not commute one cannot expect that (2.19) holds in general, i.e., for all states. That situation is related to what specialists call "breaking the classical probability law".

In the following some comments: From (2.17) and (2.18) we obtain two probability measures P_{B1,B_2} and P_{B2,B_1} on R^2 related to two random vectors each of them consisting of two components. In both cases we can define the following (classical) conditional probabilities

(a) "distribution of second component under condition we know the first component"

(b) "distribution of first component under condition we know the second component"

The conditional distributions (a) make sense from the practical point of view. It represents the behaviour of the output of the second measurement under condition that we know the output of the first measurement. The conditional distribution of the first measurement under condition that we know the output of the second measurement (case (b)) may be senseless because the first measurement has already been done.

Furthermore, one has to take into account that P_{B1,B_2} and P_{B2,B_1} describe two different procedures. For that reason it seems to be obvious that using conditional probabilities calculated according to P_{B1,B_2} for calculations based on P_{B2,B_1} will cause some trouble, i.e., basic formulas from classical probability will be violated if $P_{B1,B2} \neq P_{B2,B_1}$.

Nevertheless, from the mathematical point of view it is interesting to check whether one has joint probabilities because in that case one can apply formulas from classical probability like Bayes formula.

ON THE QUANTUM ZENO EFFECT

Starting with the notations used in the previous section we will specialize

$$B_k = B := \sum_{j=1}^{m} z_j \, Pr_{H_j}, \quad k = 1,\dots,n \tag{3.1}$$

$$t_k := \frac{k-1}{n} T \quad k = 1,\dots,n \tag{3.2}$$

That means we consider repeated measurements in the time interval $[0,T]$ according to B starting at time $t_1 = t_0 = 0$ with distance of time

$$\tau_k = \tau := \frac{1}{n} T, \quad k = 2,\dots n. \tag{3.3}$$

Further we consider the following semi group of unitary operators on H

$$V_t := e^{-ith}, t \geq 0 \tag{3.4}$$

where the selfadjoint operator h is given by its unique representation

$$h := \sum_{j=1}^{r} c_j Pr_{D_j} \tag{3.5}$$

i.e., $(D_j)_{j=1}^{r}$ represents an orthogonal decomposition of

H and $c_j \neq c_s$ if $j \neq s$

Again ρ denotes the state of the quantum system at time 0, and the random numbers $\eta_1,...\eta_n$ represent the outputs of measurements.

Then we obtain the following theorem.

Theorem 3 it holds

$$P(\eta_k = \eta_1; k = 2,...,n) \xrightarrow[n \to +\infty]{} 1 \tag{3.6}$$

Immediately from theorem 3 we get the following corollary.

Corollary 1

$$P(\eta_k = z_j; k = 1,2,...,n) \xrightarrow[n \to \infty]{} P(\eta_1 = z_j)$$
$$= tr\left(\rho Pr_{H_j}\right), j = 1,...,m. \tag{3.7}$$

Proof of Theorem 3:

We have

$$P(\eta_k = \eta_1; k = 2,...,n) = 1 - \sum_{k=2}^{n} P(\eta_k \neq \eta_1, \eta_{k-1} = \eta_1, ..., \eta_2 = \eta_1) = 1\sum_{k=2}^{n}\sum_{j=1}^{m} P(\eta_k \neq z_j, \eta_{k-1} = z_j, ...\eta_1 = z_j) \tag{3.8}$$

For that reason (3.6) is equivalent to

$$\sum_{k=2}^{n} P\left(\eta_k \neq z_j, \eta_s = z_j; s = 1,\ldots,k-1\right) \xrightarrow[n\to\infty]{} 0,$$
$$j = 1,\ldots,m. \tag{3.9}$$

Now we fix $k(=2,\ldots n), j(=1,\ldots m)$, and define indicator functions putting

$$f_s := \begin{cases} \chi_{\{z_j\}}, & s = 1,\cdots,k-1, \\ \chi_{\{y\in R: y\neq z_j\}}, & s = k, \\ 1, & s = k+1,\cdots,n. \end{cases} \tag{3.10}$$

Then we get (cf. (2.9))

$$A_{B,t_1,f_1,\cdots,B,t_n,f_n}$$
$$= V_\tau^* \left[\left(\left(V_\tau \Pr_{H_j} \right)^{k-1} \right)^* \left(I_H - \Pr_{H_j} \right) \left(V_\tau \Pr_{H_j} \right)^{k-1} \right] V_\tau. \tag{3.11}$$

Further because of Theorem 2 we have

$$P\left(\eta_k \neq z_{j,} \eta_{k-1} = z_j, \ldots, \eta_1 = z_j\right)$$
$$[0,T[= \operatorname{tr}\left(\rho A_{B,t_1,f_1,\ldots B,t_n,f_n}\right). \tag{3.12}$$

Proving Theorem 3, we can restrict ourselves to the case of a pure state ρ related to $x \in H, \|x\| = 1,$.

Then using (3.11) and (3.12) we obtain

$$P\left(\eta_k \neq z_j, \eta_{k-1} = z_j, \cdots, \eta_1 = z_j\right) = \left\langle \left(V_\tau \Pr_{H_j}\right)^{k-1} V_\tau x \middle| \left(I_H - \Pr_{H_j}\right)\left(V_\tau \Pr_{H_j}\right)^{k-1} V_\tau x \right\rangle$$
$$= \left\langle \left(V_\tau \Pr_{H_j}\right)^{k-2} V_\tau x \middle| \Pr_{H_j} V_\tau^* \left(I_H - \Pr_{H_j}\right) V_\tau \Pr_{H_j} \left(V_\tau \Pr_{H_j}\right)^{k-2} V_\tau x \right\rangle. \tag{3.13}$$

Using (3.4) and (3.5) we get

$$V_\tau^*\left(I_H - \Pr_{H_j}\right)V_\tau = I_H - V_\tau^* \Pr_{H_j} V_\tau = \sum_{s_1, s_2}^{r}\left(1 - \mathrm{e}^{i\tau\left(c_{s_1} - c_{s_2}\right)}\right)\Pr_{D_{s_1}} \Pr_{H_j} \Pr_{D_{s_2}} + I_H - \Pr_{H_j}.$$

That implies

$$\Pr_{H_j} V\tau^*\left(I_H - \Pr_{H_j}\right)V_\tau \Pr_{H_j} = \Pr_{H_j}\left[\sum_{s_1, s_2}^{r}\left(1 - \mathrm{e}^{i\tau\left(c_{s_1} - c_{s_2}\right)}\right)\Pr_{D_{s_1}} \Pr_{H_j} \Pr_{D_{s_2}}\right]\Pr_{H_j}$$
$$= -\Pr_{H_j}\left[\sum_{s_1, s_2}^{r}\left(\sum_{l \geq 2} \frac{\left[i\tau\left(c_{s_1} - c_{s_2}\right)\right]^l}{l!}\right)\Pr_{D_{s_1}} \Pr_{H_j} \Pr_{D_{s_2}}\right]\Pr_{H_j}$$
$$- \Pr_{H_j}\left[\sum_{s_1, s_2}^{r} i\tau\left(c_{s_1} - c_{s_2}\right)\Pr_{D_{s_1}} \Pr_{H_j} \Pr_{D_{s_2}}\right]\Pr_{H_j}. \tag{3.14}$$

Easy calculations show

$$\Pr_{H_j}\left[\sum_{s_1, s_2}^{r} i\tau\left(c_{s_1} - c_{s_2}\right)\Pr_{D_{s_1}} \Pr_{H_j} \Pr_{D_{s_2}}\right]\Pr_{H_j} = i\tau\left(\Pr_{H_j} h \Pr_{H_j} I_H \Pr_{H_j} - \Pr_{H_j} I_H \Pr_{H_j} h \Pr_{H_j}\right) = 0. \tag{3.15}$$

Because of $\|\mathrm{x}\| = 1$ we obtain using the Schwartz inequality

$$\left\langle \left(\mathrm{V}_\tau \mathrm{Pr}_{\mathrm{H_j}}\right)^{\mathrm{k-2}} \mathrm{V}_\tau \mathrm{x} \middle| \mathrm{Pr}_{\mathrm{H_j}} \mathrm{Pr}_{\mathrm{D_{s1}}} \mathrm{Pr}_{\mathrm{H_j}} \mathrm{Pr}_{\mathrm{D_{s2}}} \mathrm{Pr}_{\mathrm{H_j}} \left(\mathrm{V}_\tau \mathrm{Pr}_{\mathrm{H_j}}\right)^{\mathrm{k-2}} \mathrm{V}_\tau \mathrm{x} \right\rangle \leq 1 \tag{3.16}$$

Now using (3.3), (3.13), (3.14), (3.15) and (3.16) we can estimate

$$\sum_{k=2}^{n} P\left(\eta_k \neq z_j, \eta_s = z_j; s=1,\cdots,k-1\right) \leq \frac{T^2}{n} \sum_{s_1,s_2}^{r} \left(c_{s_1} - c_{s_2}\right)^2 e^{\frac{T}{n}\left|c_{s_1} - c_{s_2}\right|}, \; j=1,\cdots,m. \tag{3.17}$$

That implies (3.9) what proves Theorem 3.

Remark 3 Using (3.8) and (3.17) we obtain

$$1 - P(\eta_k = \eta_1; k=2,\ldots n)$$
$$\leq \frac{T^2 m}{n} \sum_{s_1,s_2}^{r} \left(c_{s_1} - c_{s_2}\right)^2 e^{\frac{T}{n}\left|c_{s_1} - c_{s_2}\right|}$$

Thus for given $\varepsilon > 0$ we can calculate n_ε such that always holds

$$P(\eta_k = \eta_1; k=2,\ldots,n) \geq 1-\varepsilon \quad (n \geq n_\varepsilon)$$

REFERENCES

1. E. C. G. Sudarshan and B. Misra, "The Zeno's Paradox in Quantum Theory," Journal of Mathematical Physics, Vol. 18, No. 4, 1977, pp. 756-763.
2. T. Nakanishi, K. Yamane and M. Kitano, "AbsorptionFree Optical Control of Spin Systems: The Quantum Zeno Effect in Optical Pumping," Physical Review A, Vol. 65, No. 1, 2001, Article ID: 013404. http://dx.doi.org/10.1103/PhysRevA.65.013404
3. P. Facchi, D. A. Lidar and S. Pascazio, "Unification of Dynamical Decoupling and the Quantum Zeno Effect," Physical Review A, Vol. 69, No. 3, 2004, Article ID: 032314.http://dx.doi.org/10.1103/PhysRevA.69.032314
4. A. Degasperis, L. Fonda and G. C. Ghirardi, "Does the Lifetime of an Unstable System Depend on the Measuring Apparatus?" Il Nuovo Cimento A Series 11, Vol. 21, No. 3, 1974, pp. 471-484. http://dx.doi.org/10.1007/BF02731351
5. C. Teuscher and D. Hofstadter, "Alan Turing: Life and Legacy of a Great Thinker," Springer, Berlin, Heidelberg, 2004. http://dx.doi.org/10.1007/978-3-662-05642-4
6. J. von Neumann, "Mathematische Grundlagen der Quantenmechanik," Springer, Berlin, Heidelberg, 1932.
7. K.-H. Fichtner and L. Fichtner, "Bosons and a quantum model of the brain, Jenaer Schriften zur Mathematik und Informatik," Math/Inf/08/05, Faculty of Mathematics and Informatics, FSU Jena, Jena, 2005, 27 p.

8. K.-H. Fichtner and L. Fichtner, "Quantum Models of Brain Activities I—Recognition of Signals," In: J. C. Garcia, R. Quezada and S. B. Sontz, Eds., Quantum Probability and Related topics, XXIII of QP-PQ: Quantum Probability and White Noise Analysis, World Scientific, New Jersey, London, Singapore, 2008, pp. 135-144.
9. K.-H. Fichtner, L. Fichtner, W. Freudenberg and M. Ohya, "On a Mathematical Model of Brain Activities," Quantum Theory, Reconsideration of Foundations-4, 962 of AIP Conference Proceedings, American Institute of Physics, Melville, New York, 2007, pp. 85-90.
10. K.-H. Fichtner, L. Fichtner, W. Freudenberg and M. Ohya, "On a Quantum Model of the Recognition Process," In: L. Accardi, W. Freudenberg and M. Ohya, Eds., Quantum Bio-Informatics, XXI of QP-PQ: Quantum Probability and White Noise Analysis, World Scientific, New Jersey, London, Singapore, 2008, pp. 64-84.
11. K.-H. Fichtner, L. Fichtner, W. Freudenberg and M. Ohya, "On a Quantum Model of the Brain Activities," In: L. Accardi, W. Freudenberg and M. Ohya, Eds., Quantum Bio-Informatics III, XXVI of QP-PQ: Quantum Probability and White Noise Analysis, World Scientific, New Jersey, London, Singapore, 2010, pp. 81-92.
12. K.-H. Fichtner, L. Fichtner, W. Freudenberg and M. Ohya, "Quantum Models of the Recognition Process—On a Convergence Theorem," Open Systems and Information Dynamics, Vol. 17, No. 2, 2010, pp. 161-187.http://dx.doi.org/10.1142/S1230161210000114
13. K.-H. Fichtner, L. Fichtner, W. Freudenberg and M. Ohya, "Self-Collapses of Quantum Systems and Brain Activities," In: L. Accardi, W. Freudenberg and M. Ohya, Eds., Quantum Bio-Informatics IV, XXVIII of QP-PQ: Quantum Probability and White Noise Analysis, World Scientific, New Jersey, London, Singapore, 2011, pp. 101-115.
14. K.-H. Fichtner, L. Fichtner, K. Inoue and M. Ohya, "Internal Noise Caused by the Memory," Open Systems and Information Dynamics, Vol. 18, No. 4, 2011, pp. 405- 422.http://dx.doi.org/10.1142/S1230161211000285
15. K.-H. Fichtner, K. Inoue and M. Ohya, "On a Quantum Model of Brain Activities—Distribution of the Outcomes of EEG-Measurements and Random Point Fields," Open Systems and Information Dynamics, Vol. 19, No. 4, 2012, Article ID: 1250025.
16. J. M. Schwartz, H. P. Stapp and M. Beauregard, "Quantum Physics in Neuroscience and Psychology: A Neurophysical Model of Mind-Brain Interaction," Philosophical Transactions of the Royal Society of London B, Vol. 360, No. 1458, 2005, pp. 1309-1327.http://dx.doi.org/10.1098/rstb.2004.1598
17. R. Penrose, "The Emperor's New Mind: Concerning Computers, Minds and The Laws of Physics," Oxford University Press, Oxford, 1989.
18. S. Hameroff and R. Penrose, "Orchestrated Objective Reduction of Quantum Coherence N Brain Microtubules: The 'Orch OR' Model Forconsciousness," Mathematics and Computer Simulation, Vol. 40, No. 3-4, 1996, pp. 453-480. http://dx.doi.org/10.1016/0378-4754(96)80476-9

19. R. Hari and O. V. Lounasmaa, "Neuromagnetism: Tracking the Dynamics of the Brain," Physics World, Vol. 13, 2000, pp. 33-38.
20. W. Tirsch, "Biomedizinische Relevanz der quantitativen EEG Analyse," LMU München, München, 2009.
21. K.-H. Fichtner, "Time Series Related to Quantum Measurements and the Quantum Zenon Effect," Jenaer Schriften zur Mathematik und Informatik, Math/Inf/02/2012, Faculty of Mathematics and Informatics, FSU Jena, Jena, 2012, 15 p.
22. M. Asano, M. Ohya, Y. Tanaka, A. Khrennikov and I. Basieva, "Quantum Like Representation of Baysian Updating," American Institute of Physics, Vol. 1327, No. 1, 2011, pp. 57-62.
23. M. Asano, I. Basieva, A. Khrennikov, M. Ohya and I. Yamato, "A General Quantum Information Model for the Contextual Dependent Systems Breaking the Classical Probability Law," ArXiv:1105.4769v1(quant-ph), 21 May 2011.

CITATION

K. Fichtner and K. Inoue, "On the Quantum Zeno Effect and Time Series Related to Quantum Measurements," Applied Mathematics, Vol. 4 No. 10C, 2013, pp. 61-69. doi: 10.4236/am.2013.410A3008.

ARIMA: An Applied Time Series Forecasting Model for the Bovespa Stock Index

Paulo Rotela Junior, Fernando Luiz Riêra Salomon, and Edson de Oliveira Pamplona
Institute of Production Engineering and Management, Federal University of Itajuba, Itajuba, Brazil

4

ABSTRACT

Due to the relative uncertainty involved with the variables which affect financial market behavior, forecasting future variations in a time series of the Brazilian stock market Index (Ibovespa) can be considered a difficult task. This article aims to evaluate the performance of the model ARIMA for time series forecasting of Ibovespa. The research method utilized was mathematical modeling and followed the Box-Jenkins method. In order to compare results with other smoothing models, the parameter of evaluation MAPE (Mean Absolute Percentage Error) was used. The results showed that the model utilized obtained lower MAPE values, thus indicating greater suitability. This therefore demonstrates that the ARIMA model can be used for time-series indices related to stock market index forecasting.

INTRODUCTION

Economic crises in recent decades and the consequent financial losses demonstrate that markets, financial institutions and investors urgently needed to improve their models to measure and predict the risks to which they were exposed. Equity investments become a great alternative when compared to other applications, especially in long periods [1].

Predicting the future behavior of a time series of data on the Bovespa Index (Index of shares of the São Paulo Stock Exchange) is not an easy task, given the uncertainties related to the variables that affect the behavior of financial markets and how they will impact prices practiced in the future. Studies applied to forecasting financial time series of assets, indices and investment portfolios are an important tool used for decision making in many areas, investment managers, asset pricing, and the areas responsible for risk management.

The genius of the work of Markowitz [2] in proving that the risk of a portfolio consisting of n assets is less than the weighted sum of the risks of each asset, given a different correlation of 1, laid the foundations of portfolio selection with a focus not only profitability, but the relationship between the risk taken and their return. This study, along with the work of Sharpe [3] and Fama [4] originated Modern Financial Theory, which combines the risk taken to the expected return in an efficient relationship.

This paper aims to evaluate the performance of the ARIMA model to predict the time series of the Bovespa Index, measured by MAPE (mean absolute error percentage) and compare it with other models. Historical data of monthly Bovespa quotations from January 1995 to January 2013 were used. The models were used to compare Single Exponential Smoothing and Double Exponential Smoothing.

This paper is organized as follows: In Section 2, a review of forecasting is presented. In Section 3, the research method is shown. In Section 4, analysis of historical data of the Bovespa Index, data transformation, the necessary adjustments and calculation of MAPE values are all examined based on the results found. Section 5 contains the conclusions of the study.

FORECASTING

There are two main types of approaches to demand forecasting: qualitative methods and quantitative methods. The combination of qualitative and quantitative methods is approaching the ideal time to make a good forecast demand [5].

The main qualitative methods are: Panel data approach, Delphi method, scenario planning, educated guess, executive committee consensus, sales force survey, Historical Analysis and Market Research [5] -[8].

Quantitative methods are based on historical data (time series) and assume that past results are relevant for predicting the future [7]. The classical methods of time series are: Moving Average, Exponential Adjustment, Linear Trend, and Nonlinear Trend. These methods require that the series is stationary, i.e. the mean and covariance are constant between the periods. In this context, the auto regressive methods for stationary series, and AR (Auto Regressive) and ARMA (Auto Regressive Move Average) are the most suitable because they generate a more reliable prediction [9].

Another critical effect on time series is the presence of seasonality, i.e. oscillations or disturbances in series occurring at regular intervals of less than one year. And, according to Bacci [8] , quantitative ARIMA models can describe two classes of processes: linear stationary processes and linear homogeneous non-stationary processes. The stationary linear processes use basically three types of models: auto regressive model of order p(AR(p)); moving averages of order q (MA(q)) and auto regressive and moving average process of order p and q(ARMA(p, q)). Non-homogeneous stationary linear processes assume that the series are non-stationary in level and/or slope [10]. According to Pindyck and Rubinfeld [11] , the amount of times that the original series has to be differentiated to result in a stationary series is called order of homogeneity.

Some stationary random processes (forward constant average over time) can be modeled by means of a mixed autoregressive process and moving average ARMA(p, q). Makridakis, Wheelwrigt and Hyndman [12] state that in this case, depending on the values of p and q, Y_t will depend on the past p values of Y and past q values of Θ errors. This process is Equation (1):

$$Y_t = \varphi_1 Y_{t-1} + \varphi_2 Y_{t-2} + \cdots + \varphi_p Y_{t-p} + \delta + \varepsilon_t - \Theta_1 \varepsilon_{t-1} - \Theta_1 \varepsilon_{t-2} - \cdots - \Theta_q \varepsilon_{t-q} \,. \quad (1)$$

For the process shown in (1), the stationary sum $\varnothing_1 + \varnothing_2 + \ldots + \varnothing_p$ should be less than 1 [11]. According to Pindyck and Rubinfeld [11] and Fava [13] , non-stationary series can be transformed into stationary series where observations are differentiated one or more times.

The first differentiation of the data is in Equation (2):

$$Z_t = \Delta Y_t = Y_t - Y_{t-1}. \tag{2}$$

Being:

Y_t = observation Y, in period t of the series Y_t without differentiation;

Y_{t-1} = observation Y, in period t − 1 of the series Y_t without differentiation;

$\Delta Y_t = Z_t$ = observation Z, in period t, belonging to series Z_t with data from the series Y_t differentiated for the first time.

The data series will be differentiated for the first time in the following manner: the value of the second datum is decreased from the first; the third will be decreased in the second, the fourth from the third, and so on. With this process, the differentiated series for the first time, Z_t will have one less observation (n − 1 observations) than the original series Y_t.

The second differentiation of data can be represented by Equation (3):

$$W_t = \Delta^2 Y_t = \Delta\left(\Delta Y_t\right) = \Delta\left(Y_t - Y_{t-1}\right). \tag{3}$$

According to Bacci [8] , the series Y_t differentiated a second time, or series Z_t diffentiated once, will lead to the series W_t.

The differential data series Z_t is obtained as follows: the value of the second observation decreased from the first observation forms the first observation, the value of the third diminished from the second provides the second and so on.

The twice differentiated series (W_t) will have n − 2 observations compared to the original series Y_t. Thus, after one or more differentiations of Y_t series to make it stationary, it produces a series stationary W_t, which can now be modeled as an ARMA process (p,q).

According to Pindyck and Rubinfeld [11] , the initial series Y_t is an autoregressive process of order (p,d,q) integrated moving average (MA), given by Equation (4):

$$\phi(B)\Delta^d Y_t = \delta + \theta(B)\varepsilon_t . \tag{4}$$

Being:

$\Delta^d Y_t = W_t$; with d = order of the stationary series W_t, that is, the number of times that the non-stationary series Y_t was differentiated until becoming a stationary series W_t;

ϕ (B)=autoregressive oerator;

θ (B)=moving average operator;

where operator B imposes a time delay of one period each time it is applied to a variable Y_t.

The construction of an ARIMA model is based on a cycle with the following stages [10] : Identification of a general class of models which will be analyzed; Specification of the model, based on autocorrelation analysis, partial autocorrelations and other criteria; Estimation of model parameters; Verification of the adjusted model, which is done by means of residual analysis to measure its suitability to carry out the forecast; If the model is not suitable, the cycle repeats from the identification of the model.

MATERIALS AND METHODS

According to Bertrand and Fransoo [14] , this investigation can be classified as applied, having a descriptive empirical goal, since the researchers are interested in creating a model that adequately describes the causal relationships which may exist in reality, leading to the understanding of the current processes thus, fostering understanding of real processes. The approach to the problem was quantitative, and a mathematical modeling research method adopted. Figure 1 shows the Box-Jenkins method [15] :

Initially data collection was done using Economatica® software, using monthly closing prices of the Bovespa Index, for the period January 2000 to December 2012.

RESULTS ANALYSIS

The values obtained were plotted using Minitab® 16 Statistical Software for an initial evaluation of the data as shown in Figure 2.

It can be seen that the data is not stationary and the series presents variance from one period to another. The analysis used in this series demonstrated the need for a logarithmic transformation on the data which generate Figure 3. Figure 4 and Figure 5 show ACF (autocorrelation) and PACF (partial autocorrelation) transformed time series.

Tests for ACF (autocorrelation) and PACF (partial autocorrelation) indicated that the AR1 model ARIMA (0, 2, 1) model could be used to predict the behavior of the series, shown in Table 1 andTable 2.

A verification of the series' residuals transformed by Log Bovespa Index and both tests were carried out, through which it was demonstrated that autocorrelation does not exist between series residuals, which enabled the utilization of both to forecast the behavior of the series, as shown in Figures 6-9.

The next step was to realize tests to verify the accuracy of the models. Initially the model was used to forecast 10 months ahead in several periods of the series, trying to compare the MAPE between these periods, as shown in Figure 10 and Figure 11.

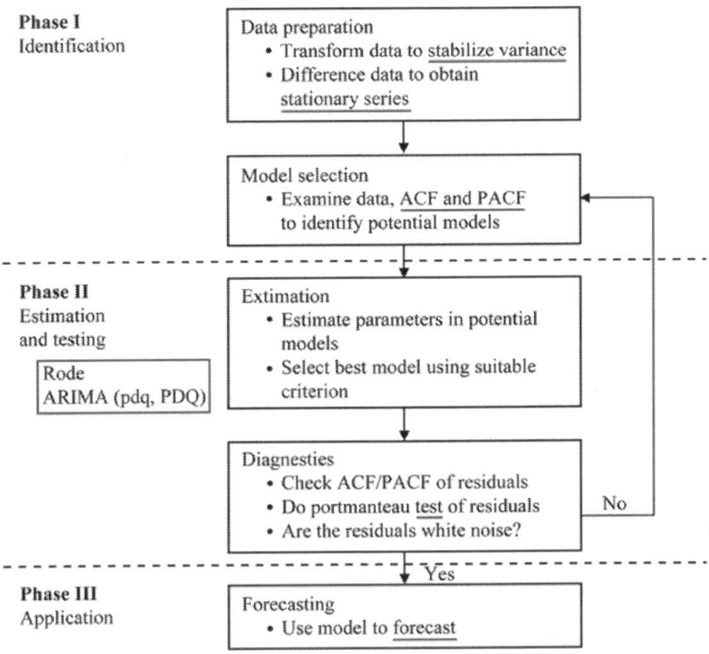

Figure 1: Box-Jenkins Methodology. Source: Makridakis, Wheelwrigt and McGee [15].

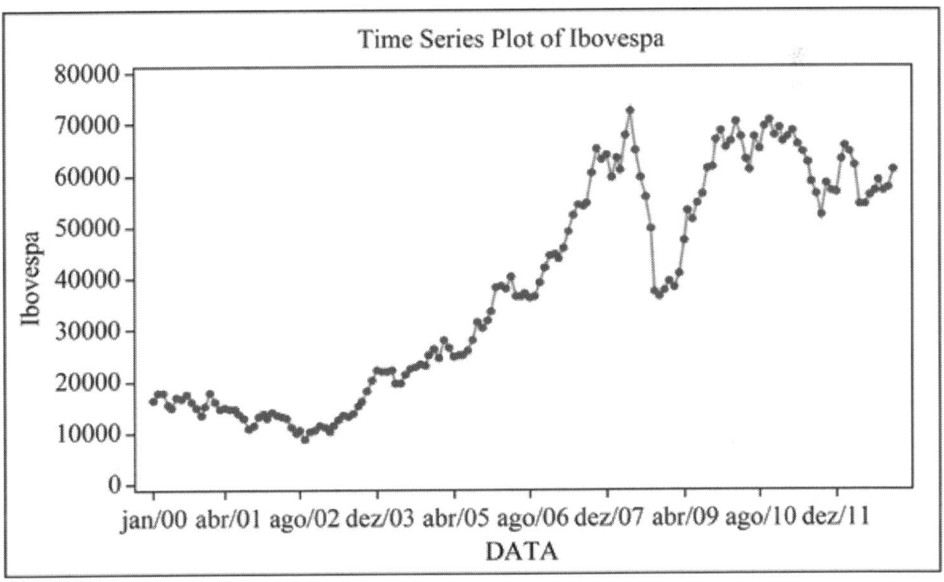

Figure 2: Plotting the Ibovespa time series.

Table 1: Coefficient statistics of the AR1 model

	Coefficient	Coefficient standard error	T statistic	P-value
AR 1	0.9996	0.0006	1697.01	0.000

Table 2: Chi-square statistics for the Box-Pierce modification (Lhung-Box)

Lag	12	24	36	48
Chi-square	11.5	18.5	26.2	33.5
Degrees of freedom	11	23	35	47
P-value	0.399	0.728	0.858	0.931

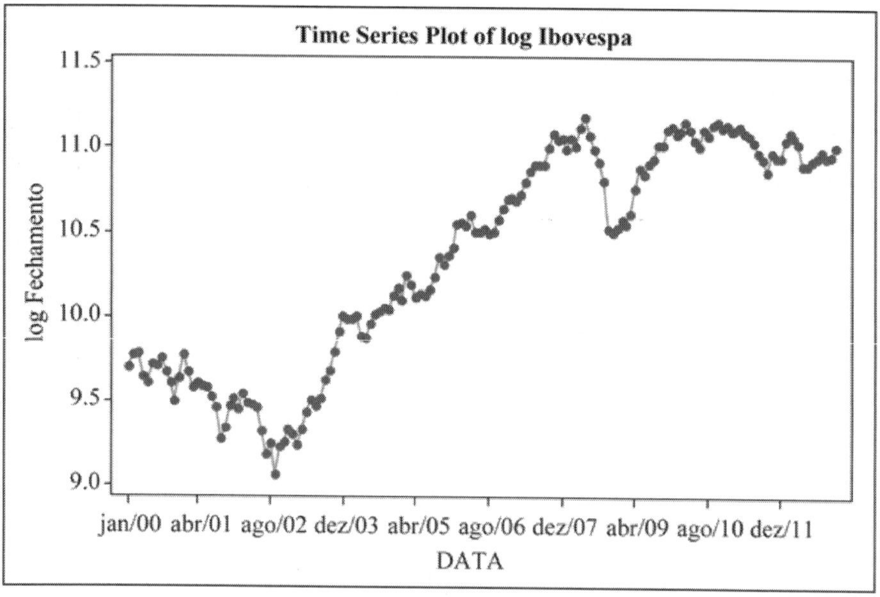

Figure 3: Logarithmic plotting of the Ibovespa time series.

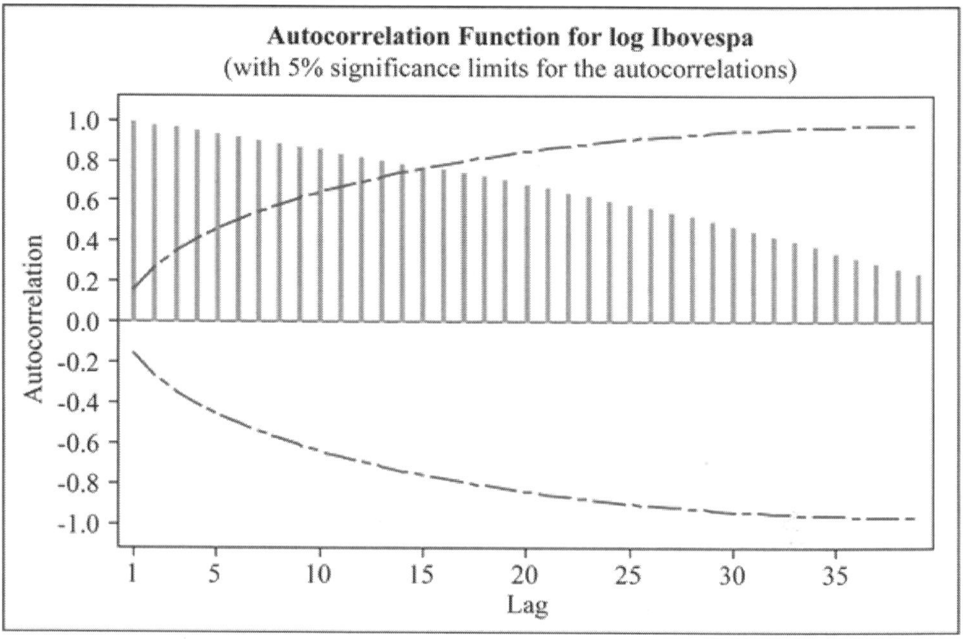

Figure 4: ACF of logarithmic Ibovespa.

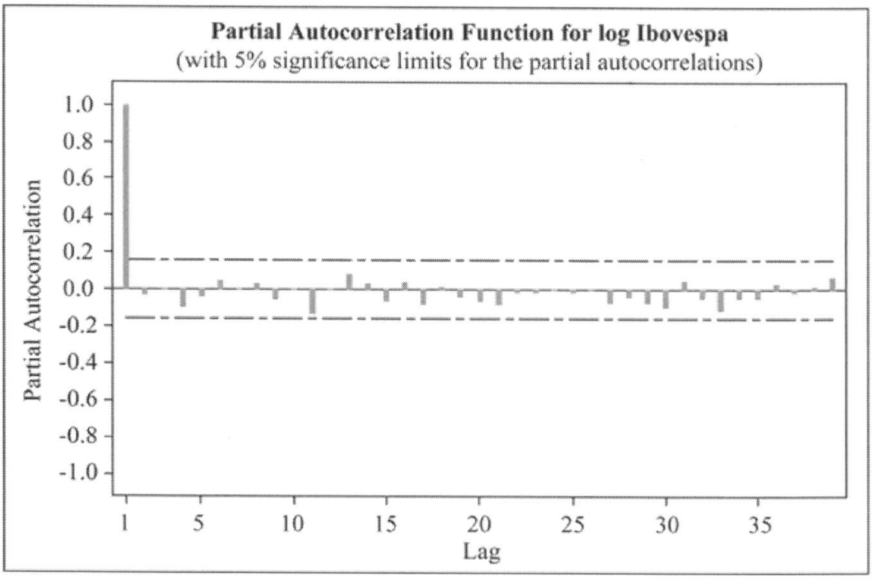

Figure 5: PACF of logarithmic Ibovespa.

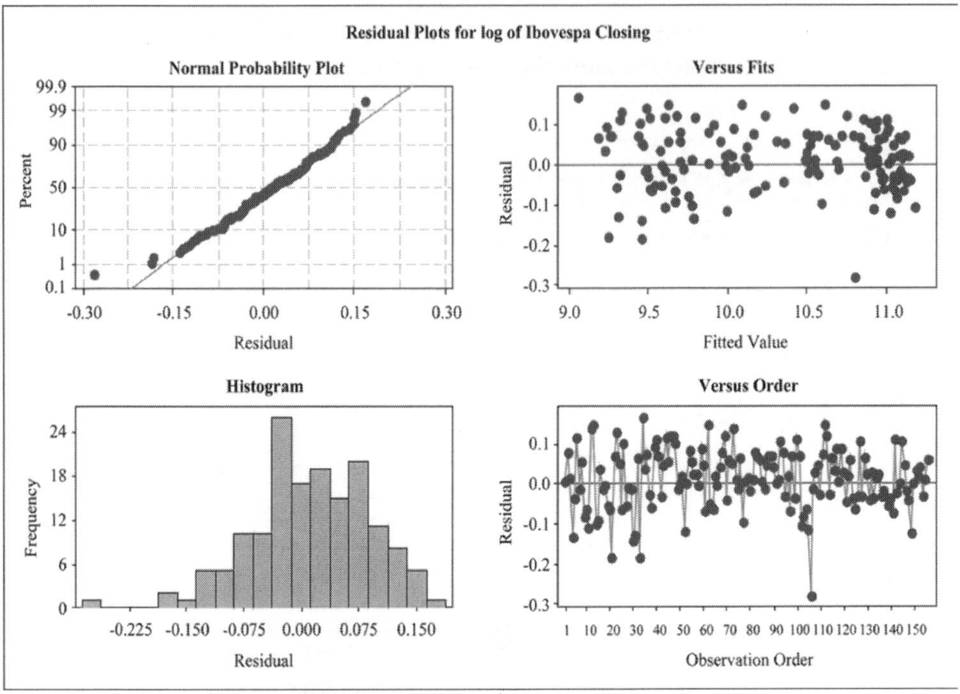

Figure 6: Residual plotting.

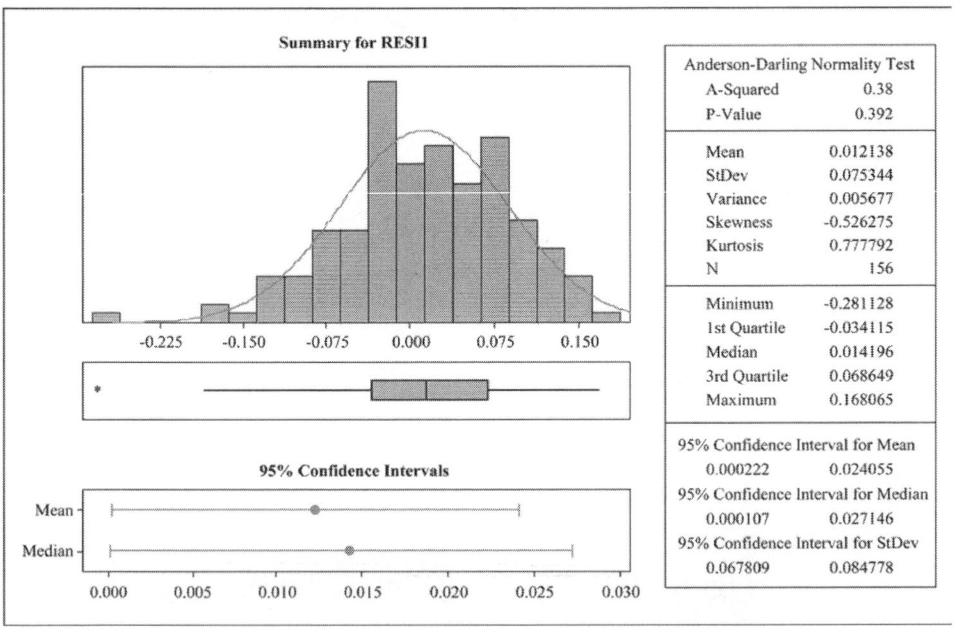

Figure 7: Graphical summary of the residues.

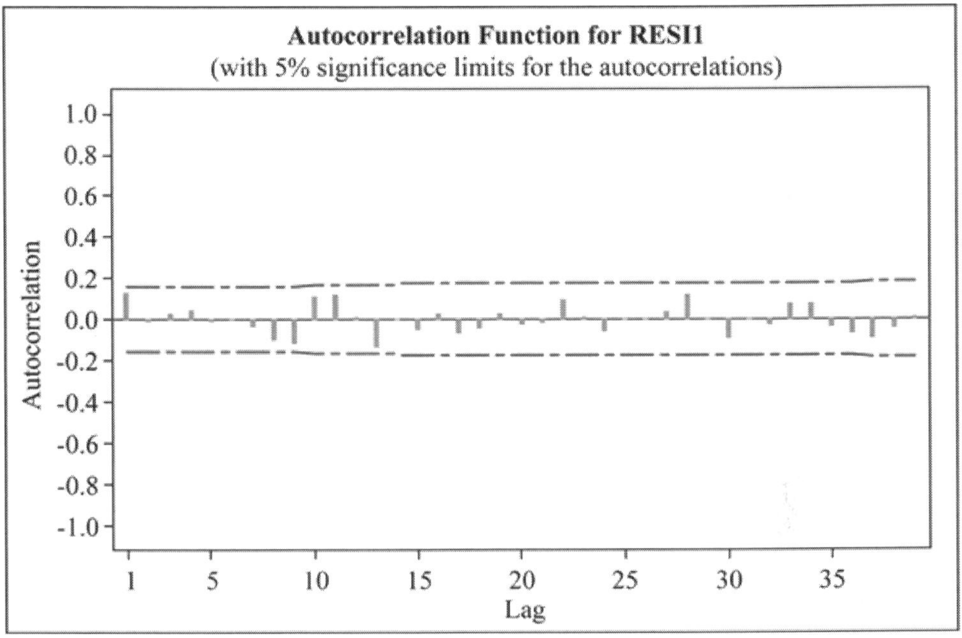

Figure 8: ACF of residuals.

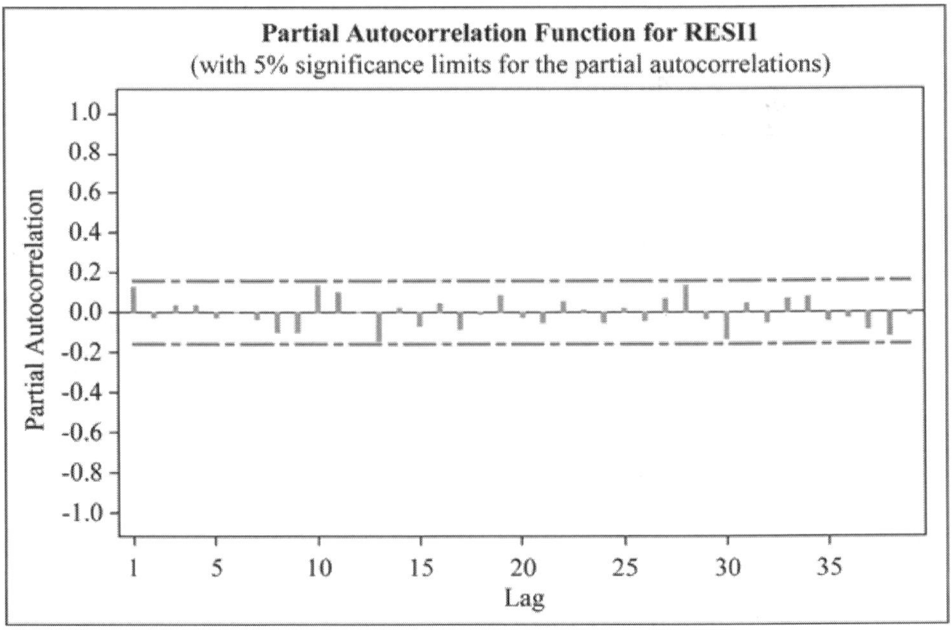

Figure 9: PACF of residuals.

The use of this indicator (MAPE) to evaluate the models was used because this measures the absolute average, i.e., the sum of the percentage errors, in which the series data values undergo alterations throughout time, which influences the size of the error. Using the value of the error divided by the value of the observation, transforming the error as a percentage of this observation, diminishes the effects caused by variation of the values of the series, allowing one to compare the error between observations of distinct values.

A test of forecasting 10 months ahead in five distinct periods showed that, in all models, the error tends to increase after the second period, significantly impacting the average error. This fact could result from the choice of one model over another simply because of the errors from the second period are lower. This finding could be valid, if the prediction is used to making decisions in the midterm, in which the prediction of several periods ahead would determine actions to be taken and which could hardly be changed in the short term.

In the case of financial time series, significant changes in the forecasts made can trigger immediate decisions and corrections in a matter of minutes, or at worst a few days, repositioning strategies by hedging or even the rebalancing of the portfolio by the complete elimination of certain positions that would be affected or even entire investment strategies in a short space of time.

Thus, it was chosen to perform the forecast one-step-ahead, in the case of this study, one month ahead. In the various models analyzed, this prediction proved to be such as that found the lowest MAPE. Thus, it avoids using only a period of the series, analyze the ASM one-step-ahead composed of five periods, which evaluated the forecast of each of these periods, being the MAPE constructed from the absolute percent average of the sum of these errors.

Table 3 shows the comparison of MAPE obtained by using four AR1 models, Single Exponential Smoothing, Double Exponential Smoothing and ARIMA (0, 2, 1). It is observed that the MAPE of the AR1 model enables its use when compared to other options.

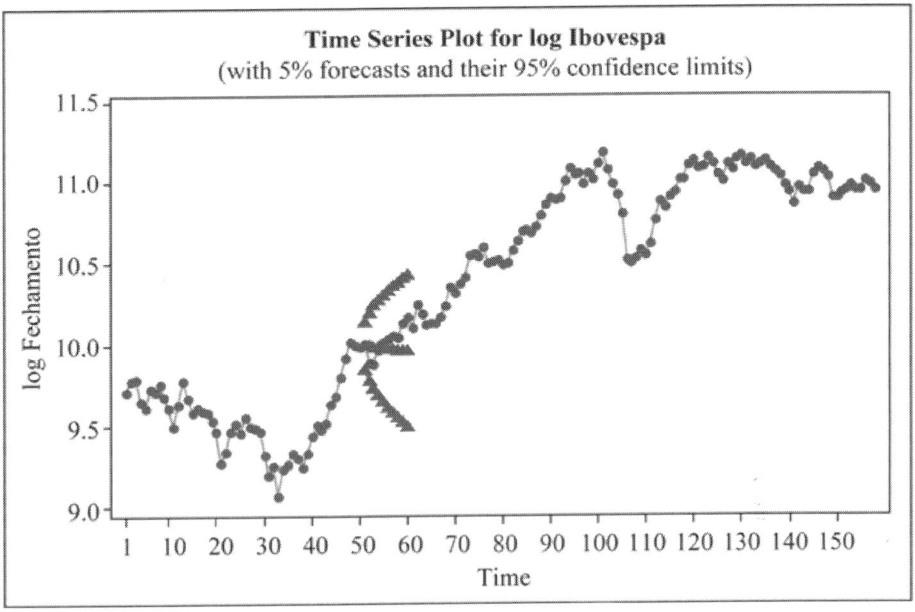

Figure 10: Forecast 10 points ahead, based on month 50.

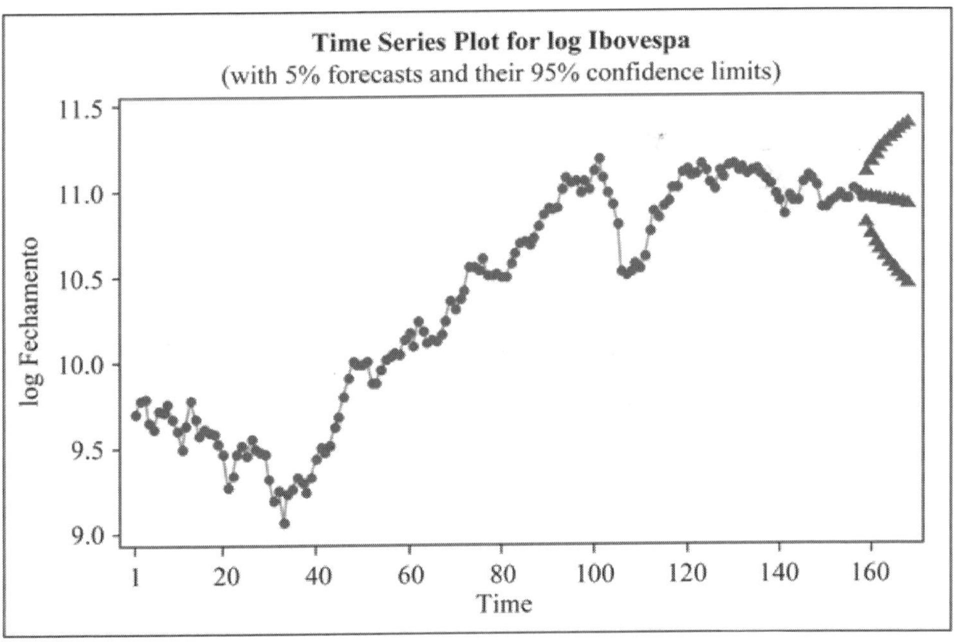

Figure 11: Forecast 10 points ahead, based on the last month.

CONCLUSIONS

Through the results obtained, it is observed that the model is effective in its forecasts. The statistics of the AR1 model coefficients and Chi-square statistics for modified Box-Pierce (Ljung-Box) provide proof to this fact.

A MAPE (mean absolute error percentage) of 0.052% was obtained, a lower value than those found in predictions made with other models used for comparison.

This study sought to obtain short-term forecasts for the next month (one step ahead) in order to minimize pre- diction errors. The model can be considered adequate for predicting the Bovespa Index series, and can be used as an aid to decision-making mechanism.

Table 3: MAPE table (absolute average percent error)

	AR1	Single Exp smoothing	Double Exp smoothing	ARIMA (0, 2, 1)
MAPE	0.052%	0.086%	0.118%	0.064%

ACKNOWLEDGEMENTS

The authors would like to express their gratitude to the Brazilian agencies CNPq (National Counsel of Technological and Scientific Development), CAPES (Post-Graduate Federal Agency), and FAPEMIG (Foundation for the Promotion of Science of the State of Minas Gerais), which have been supporting the efforts for the development of this work in different ways and periods.

REFERENCES

1. Rotela, J.P., Pamplona, E.O. and Salomon, F.R. (2014) Otimização de Portfólios: Análise de Eficiência. RAE—Revista de Administração de Empresas, 54, 405-413.http://dx.doi.org/10.1590/S0034-759020140406
2. Markowitz, H. (1952) Portfolio Selection. Journal of Finance, 7, 77-91.

3. Sharpe, W.F. (1963) A Simplified Model for Portfolio Analysis. Management Science, 9, 277-293.
4. Fama, E. (1970) Efficient Capital Markets: A Review of Theory and Empirical Work. Journal of Finance, 25, 383-417. http://dx.doi.org/10.2307/2325486
5. Slack, M., Chambers, S. and Johnston, R. (2007) Operations Management. Prentice Hall/Financial Times, Upper Saddle River.
6. Linstone, H.A. and Turoof, M. (1975) The Delphi Method: Techniques and Applications. Addison-Wesley, Boston.
7. Gaither, N. and Frazier, G. (2001) Operations Management. South-Western, Ohio.
8. Bacci, L.A. (2007) Combinação de métodos de séries temporais para a previsão da demanda de café no Brasil. Thesis, UNIFEI, Itajuba.
9. Morettin, P.A and Toloi, C.M. (2006) Análise de Séries Temporais—2ª Edição Revista e Ampliada. 2nd Edition, Editora Edgar Blucher, São Paulo.
10. Morettin, P.A. and Toloi, C.M. (1987) Previsão de séries temporais. 2nd Edition, Atual Editora, São Paulo.
11. Pindyck, R.S. and Rubinfeld, D.L. (2008) Microeconomics. 7th Edition, Prentice Hall, Upper Saddle River.
12. Makridakis, S., Wheelwrigt, S.C. and Hyndman, R.J. (1998) Forecasting: Methods and Applications. 3rd Edition, John Wiley & Sons, New York.
13. Fava, V.L. (2000) Metodologia de Box-Jenkins para Modelos Univariados. In: Vasconcellos, M.A.S. and Alves, D., Eds., Manual de econometria: nível intermediário, Atlas, São Paulo, 205-232.
14. Bertrand, J. and Fransoo, J. (2002) Operations Management Research Methodologies Using Quantitative Modeling. International Journal of Operations & Production Management, 22, 241-264. http://dx.doi.org/10.1108/01443570210414338
15. Makridakis, S., Wheelwrigt, S.C. and Mcgee, V. (1983) Forecasting: Methods and Applications. 2nd Edition, John Wiley & Sons, New York.

CITATION

Rotela Junior, P. , Salomon, F. and de Oliveira Pamplona, E. (2014) ARIMA: An Applied Time Series Forecasting Model for the Bovespa Stock Index. Applied Mathematics, 5, 3383-3391. doi: 10.4236/am.2014.521315.

Testing For Deterministic Components In Vector Seasonal Time Series

José Luis Gallego, Carlos Díaz
Departamento de Economía, Universidad de Cantabria, Santander, Spain

5

ABSTRACT

Certain locally optimal tests for deterministic components in vector time series have associated sampling distributions determined by a linear combination of Beta variates. Such distributions are nonstandard and must be tabulated by Monte Carlo simulation. In this paper, we provide closed form expressions for the mean and variance of several multivariate test statistics, moments that can be used to approximate unknown distributions. In particular, we find that the two-moment Inverse Gaussian approximation provides a simple and fast method to compute accurate quantiles and p-values in small and asymptotic samples. To illustrate the scope of this approximation we review some standard tests for deterministic trends and/or seasonal patterns in VARIMA and structural time series models.

INTRODUCTION

A wide class of test statistics for detecting the presence of deterministic components in univariate linear time series models can be derived following the [1] approach to the problem of testing for a scalar error covariance matrix in linear regression models with non-spherical disturbances. Members of this class are either locally best invariant (LBI) tests or LBI unbiased (LBIU) tests depending on whether the null hypothesis of a particular deterministic component is confronted with a one-sided or two-sided alternative, respectively. Some examples are the LBI test statistics for a null variance ratio or parametric stability proposed by [2-8], as well as the LBIU test statistics for non-invertibility or moving average (MA) unit roots derived by [9-12]. All these LBI and LBIU test statistics can be formulated as ratios of quadratic forms in normal variables, whose distribution functions are usually computed by numerical inversion of the corresponding characteristic functions using the [13] or [14] procedures. Moreover, their limiting distributions are related to that of the Cramèr-von Mises test statistics for goodness-of-fit derived and tabulated by [15]. Several non-parametric and parametric corrections have been proposed to cope with serially correlated errors so that the modified statistics follow the same limiting distributions.

Multivariate versions of these tests have been only derived in the framework of structural time series models by [5] and [16] based on the multivariate generalization of the [1] approach given by [17]. As in the univariate case, the limiting distributions of the multivariate test statistics are also related to the Cramèr-von Mises distribution. However, the small sample distributions are unknown and must be evaluated by Monte Carlo simulation. A similar problem arises in the analysis of cointegrated VAR models with Dickey-Fuller type tests, where procedures for easily computing p-values and quantiles has been proposed among others by [18], using a response surface approach, and [19], fitting a Gamma distribution with moments estimated from a response surface regression. Approximating the unknown distribution of the tests for deterministic components has been also suggested by [17], who found that such distributions can be expressed as linear combinations of Beta variates and gave closed form expressions for their first two moments, which involve the computation of the eigenvalues of a matrix whose order depends on the sample size. However, a tentative family has not been proposed yet.

In this paper, we use the results of [17] to derive closed form expressions for the mean and variance of several test statistics for deterministic components that depends on the sample size and avoid the computation of eigenvalues. Besides, we propose a two-moment Inverse Gaussian (IG) approximation to the distribution of a linear combination of Beta variates. To illustrate some applications of this approximation we provide seasonal extensions of the [5] tests that can be used to test for non-invertibility in vector seasonal ARIMA models, as well as to derive easily the [16] tests for deterministic seasonality at specific frequencies.

The paper is organized as follows. In Section 2, we summarize the main results of the [17] approach. In Section 3, we review some relevant test statistics for deterministic components in vector time series and give exact expressions for the first two moments. In Section 4, we describe the two-moment IG approximation and assess its accuracy. Finally, in Section 5, we conclude with some extensions.

INVARIANT TESTS FOR COVARIANCE STRUCTURES

Consider the multivariate linear regression model

$$\mathbf{Y} = \mathbf{X}\Pi + \mathbf{E}, \quad \mathbf{E} \sim N(0,\ \Omega(\delta) \otimes \Sigma) \tag{1}$$

where Y and Σ are T × m random matrices, X is a T × k fixed design matrix, Π is a k × m matrix of parameters, $\Omega(\delta)$ and S are T × T and m × m positive definite matrices, respectively, and δ is the parameter of interest determining whether or not the columns of E are i.i.d. Tdimensional errors, i.e., $\Omega(\delta_0) = I_T$. [17] found that the Locally Best Invariant

(LBI) test statistic of the null hypothesis $H_0: \delta = \delta_0$ against the one-sided alternative $H_1: \delta > \delta_0$ has the following general expression

$$L = \operatorname{tr}\left[\left(\hat{\mathbf{E}}'\hat{\mathbf{E}}\right)^{-1}\left(\hat{\mathbf{E}}'\mathbf{K}\hat{\mathbf{E}}\right)\right] \tag{2}$$

where tr is the trace operator, $\hat{E} = MY$ is the residual matrix in the ordinary least squares regression of Y on X, $M = I_T - X(X'X)^{-1} X'$ and K is the first derivative $d\,\Omega(\delta)/d\delta$ evaluated at $\delta = \delta_0$. Invariance is defined against the group of transformations $Y \rightarrow YP + XA$ for an arbitrary $k \times m$ matrix A and a positive definite $m \times m$ matrix P. Thus, without loss of generality, it can be assumed that $\Sigma = I_T$. From [17], and following [1], it can also be proved that the LBIU test statistic of $H_0: \delta = \delta_0$ against the two-sided alternative $H_1: \delta \neq \delta_0$ is given by (2) but with K being the second derivative $d^2\Omega(\delta)/d\delta^2$ evaluated at $\delta = \delta_0$.

The null distribution of L can be characterized rewriting it as

$$L = \sum_{t=1}^{T\text{-}k} \lambda_t B_t, \quad B_t = \boldsymbol{\varepsilon}_t'\left(\sum_{t=1}^{T-k} \boldsymbol{\varepsilon}_t \boldsymbol{\varepsilon}_t'\right)^{-1} \boldsymbol{\varepsilon}_t, \tag{3}$$

Where λ_t are the non-null eigenvalues of the product matrix MK, $\varepsilon_t \sim N(0, I_m)$, $B_t \sim$ Beta $(m/2, (T-k-m)/2)$ and $B_1 + \ldots + B_{T-k} = m$, see, e.g., [20] p. 540. [17] found that the first two moments of L are given by

$$E(L) = m\bar{\lambda}_1 \text{ and } V(L) = c\left[\bar{\lambda}_2 - \bar{\lambda}_1^2\right]$$

with $c = 2m(T-k-m)/[(T-k-1)(T-k+2)]$ and

$$\bar{\lambda}_r = [1/(T-k)]\sum_{k=1}^{T-k} \operatorname{tr}\left[(\mathbf{MK})^r\right]$$

In the next section, we give formulae to compute tr (MK) and $tr(MK)^2$ for some useful test statistics.

TEST STATISTICS

Multivariate Seasonal Random Walk plus Noise Model

[21] derived the LBI test statistic for a vector deterministic level in the multivariate local level model. We consider here the seasonal extension of this multivariate model given by

$$\mathbf{y}_t = \boldsymbol{\alpha}_t + \mathbf{u}_t, \quad \boldsymbol{\alpha}_t = \boldsymbol{\alpha}_{t-k} + \mathbf{v}_t, \quad t = 1, \cdots, T, \tag{4}$$

where the vector time series $y_t = (y_{1t}, \ldots, y_{mt})'$ is decomposed into the sum of a vector seasonal random walk $\alpha_t = (\alpha_{1t}, \cdots, \alpha_{mt})'$ plus a vector Gaussian white noise $u_t = (u_{1t}, \ldots, u_{mt})' \sim N(0, \Sigma)$, the vector Gaussian white noise $v_t = (v_{1t}, \ldots, v_{mt})' \sim N(0, \rho\Sigma)$ is assumed to be independent of u_t, and the parameter $\rho > 0$ quantifies the degree of stochasticity of a_t. Without loss of generality we assume that the seasonal period k is even and that the dataset is balanced, T = nk.

Defining the T × m matrices $Y = [y_1, \ldots, y_T]'$, $A = [\alpha_1, \cdots, \alpha_T]'$, $U = [u_1, \ldots, u_T]'$ and $V = [v_1, \ldots, v_T]'$, (4) can be written in matrix form as

$$\mathbf{Y} = \mathbf{A} + \mathbf{U}, \quad (\mathbf{D}_n \otimes \mathbf{I}_k)\mathbf{A} = (\mathbf{d}_n \otimes \mathbf{I}_k)\mathbf{A}_0 + \mathbf{V} \tag{5}$$

where $A_0 = [\alpha_{-k+1}, \cdots, \alpha_0]'$ is a k × m matrix of initial conditions, $\otimes$ denotes the Kronecker or tensor product, D_n is an n × n lower bidiagonal matrix with 1s on the main diagonal and −1s on the first sub-diagonal, which can be horizontally partitioned as $D_n = [d_n | \nabla_n]'$, being d_n = (1,

0, …, 0)′ and ∇_n the (n – 1) × n first-order differencing matrix. If A_0 is assumed to be fixed, it follows that (4) is a special case of (1) with

$$\mathbf{X} = \mathbf{i}_n \otimes \mathbf{I}_k, \quad \Pi = \mathbf{A}_0, \quad \Omega(\rho) = \mathbf{I}_T + \rho(\mathbf{C}_n \mathbf{C}_n' \otimes \mathbf{I}_k)$$

Where $C_n = D_n^{-1}$, $i_n = C_n d_n$ and X is a T × k matrix of seasonal dummy variables.

The LBI test statistic for testing the null hypothesis of deterministic seasonality (H_0: $\rho = 0$) against the alternative of seasonal random walk (H_1: $\rho > 0$), denoted by $RW_{m,k,n}$, is given by (2) with

$$\mathbf{K} = \mathbf{C}_n \mathbf{C}_n' \otimes \mathbf{I}_k \tag{6}$$

and $\hat{E}$ being the residual matrix in the multivariate regression of Y on the full set of k seasonal dummies. To compute the two first moments of RW_{mkn}, we find that the mean and mean-square of the eigenvalues of MK are given by

$$\bar{\lambda}_1 = (n+1)/6 \text{ and } \bar{\lambda}_2 = \bar{\lambda}_1\left(2n^2+7\right)/30$$

which suggest to correct the $RW_{m,k,n}$ test statistic by a factor depending of the sample size so that it converges to a non-degenerate limiting distribution. Some candidates are n, mn or m (n – k). It should be noted that $RW_{m,k,n}/mn$ has asymptotic mean 1/6 and variance 1/45 mk.

[21] also derived the LBI test statistic for a vector deterministic linear trend. We obtain here the LBI test statistic for a vector deterministic seasonal linear trend by including a vector seasonal drift β_t in (4)

$$\mathbf{y}_t = \boldsymbol{\alpha}_t + \mathbf{u}_t, \quad \boldsymbol{\alpha}_t = \boldsymbol{\alpha}_{t-k} + \boldsymbol{\beta}_t + \boldsymbol{\nu}_t, \quad \boldsymbol{\beta}_t = \boldsymbol{\beta}_{t-k} \tag{7}$$

whose matrix form is given by

$$\mathbf{Y} = \mathbf{A} + \mathbf{U},$$
$$(\mathbf{D}_n \otimes \mathbf{I}_k)\mathbf{A} = (\mathbf{d}_n \otimes \mathbf{I}_k)\mathbf{A}_0 + (\mathbf{i}_n \otimes \mathbf{I}_k)\mathbf{B}_0 + \mathbf{V}$$

where $B_0 = [\beta_{-k+1}, \ldots, \beta_0]'$ is a $k \times m$ fixed matrix of initial conditions for β_t If A_0 is fixed, (7) is a special case of (1) with $X = [i_n \otimes i_k, t_n \otimes i_k], \Pi = [A_0, B_0]$ and $\Omega(\rho) = I_T + \rho(C_n C_n' \otimes I_k)$, where $t_n = (1, 2, \ldots, n)'$. It is now clear that the inclusion of the vector seasonal drift only affects the mean vector of the sampling distribution of Y, $X\Pi$, but not its covariance matrix. Therefore, the LBI statistic for testing the null hypothesis of vector deterministic seasonal linear trend (H_0: $\rho = 0$) against vector seasonal drifted random walk (H_1: $\rho > 0$) in (7), say $DRW_{m,k,n}$, is computed as $RW_{m,k,n}$, being now $\hat{E}$ the residual matrix in the multivariate regression of Y on k seasonal dummies and k seasonal lineal trends. We find for $DRW_{m,k,n}$ that the mean and mean-square of the eigenvalues of MK are given by

$$\bar{\lambda}_1 = (n+2)/15 \text{ and } \bar{\lambda}_2 = \bar{\lambda}_1(11n^2 + 181)/840$$

and so $DRW_{m,k,n}$ has asymptotic mean 1/15 and variance 11/6300 mk.

We also obtain another relevant modification of the $RW_{m,k,n}$ test statistic by including the time index t as a regressor in (4)

$$\mathbf{y}_t = \boldsymbol{\alpha}_t + \boldsymbol{\beta} t + \mathbf{u}_t, \quad \boldsymbol{\alpha}_t = \boldsymbol{\alpha}_{t-k} + \mathbf{v}_t \tag{8}$$

or in matrix form,

$$\mathbf{Y} = \mathbf{A} + \boldsymbol{\beta}\mathbf{t}_T + \mathbf{U}, \quad (\mathbf{D}_n \otimes \mathbf{I}_k)\mathbf{A} = (\mathbf{d}_n \otimes \mathbf{I}_k)\mathbf{A}_0 + \mathbf{V}$$

which is a special case of (1) with $X = [i_n \otimes I_k, t_T]$, $\Pi = [A_0, \beta_0]$ and $\Omega(\rho)$ as in (7). By the same token, the LBI statistic for testing H_0: $\rho =$

0 against H_1: $\rho > 0$ in (8), say $TRW_{m,k,n}$, is computed as $RW_{m,k,n}$, being now $\hat{E}$ the residual matrix in the multivariate regression of Y on k seasonal dummies and a regular lineal trend. We find for $TRW_{m,k,n}$ that the mean and mean-square of the eigenvalues of MK are given by

$$\bar{\lambda}_1 = [(5k-3)n^2 - (5k+3)]/\{30[(n-1)k-1]\}$$

and

$$\bar{\lambda}_2 = \frac{(140k-129)n^4 + (350k-213)n^2 - (490k+234)}{12600[(n-1)k-1]}$$

It should be noted that $DRW_{1,1,n} = TRW_{1,1,n}$ is the [3] test statistic for a univariate deterministic linear trend, and that the asymptotic mean and variance of $TRW_{1,1,n}/n$ are 1/15 and 11/6300, which agree with those obtained by [3] in a rather complicated proof.

Vector Seasonal IMA $(1,1)_k$ Model

Multivariate structural model (4) can be written as a vector seasonal IMA $(1,1)_k$ process

$$(\mathbf{I}_m - \mathbf{I}_m B^k)\mathbf{y}_t = (\mathbf{I}_m - \Theta \mathbf{I}_m B^k)\mathbf{a}_t, \quad \mathbf{a}_t \sim N(0, \Omega_a) \tag{9}$$

where B is the backshift operator such that $B^k y_t = y_{t-k}$, $a_t = (a_{1t}, \ldots, a_{mt})'$, W_a is a m × m positive definite matrix, and the parameter Θ is restricted to be positive so that $\rho = (1 - \Theta)^2/\Theta > 0$. Process (9) is said to be invertible when $\Theta < 1$ and strictly non-invertible when $\Theta = 1$. In the last case, the cancellation of the matrix polynomials on both sides of the equation reveals the presence of deterministic seasonality. Noting that $\rho(\Theta) = \rho(1/\Theta)$, the one-sided testing problem H_0: $r = 0$ versus H_1: $r > 0$ is equivalent to the two-sided one H_0: $\Theta = 1$ versus H_1: $\Theta \neq 1$. Hence, the LBI test statistic $RW_{m,k,n}$ for a null variance ratio in (4) is the LBIU test statistic for strict noninvertibility in (9). Note that $RW_{1,k,n}$ is the [11] test statistic for a seasonal MA unit root.

Analogously, it can be proved that $DRW_{m,k,n}$ and $TRW_{m,k,n}$ are the LBIU test statistics of H_0: $\Theta = 1$ versus H_1: $\Theta \neq 1$ in the reduced form of (7),

$$(\mathbf{I}_m - \mathbf{I}_m B^k)^2 \mathbf{y}_t = (\mathbf{I}_m - \mathbf{I}_m B^k)(\mathbf{I}_m - \Theta \mathbf{I}_m B^k)\mathbf{a}_t$$

and (8)

$$(\mathbf{I}_m - \mathbf{I}_m B)(\mathbf{I}_m - \mathbf{I}_m B^k)\mathbf{y}_t = (\mathbf{I}_m - \mathbf{I}_m B)(\mathbf{I}_m - \Theta \mathbf{I}_m B^k)\mathbf{a}_t$$

respectively. Note that $DWR_{1,k,n}$ is closely related to the [22] test statistic, while $TRW_{1,k,n}$ is the [12] test statistic for non-invertibility in the seasonal $IMA(1,1)_k$ model.

Dynamic Seasonal Linear Models

[6] considered testing the null hypothesis of deterministic seasonality against the alternative of mixed deterministic and stochastic seasonality. A related multivariate LBI test statistic can be derived in the seasonal linear regression

$$\mathbf{y}_t = \sum_{j=1}^{k} x_{jt}\boldsymbol{\beta}_{jt} + \mathbf{u}_t, \quad \boldsymbol{\beta}_{jt} = \boldsymbol{\beta}_{jt-1} + \delta_j \mathbf{v}_{jt} \tag{10}$$

where x_{jt} ($j = 1, \ldots, k$) are a full set of seasonal dummy variables, β_{jt} is either a time-varying parameter if $d_j = 1$ or a nuisance constant parameter if $d_j = 0$, $u_t \sim N(0, S)$, $v_{jt} \sim N(0, (r/k)S)$ are mutually and serially uncorrelated vector errors, and is here divided by k for comparison purposes given that (10) reduces to (4) when $d_1 + \ldots + d_K = k$. Assuming that the initial conditions $\beta_{1,0}, \ldots, \beta_{k,0}$ are fixed, (10) is a special case of (1) with $X = [x_1, \ldots, x_k]'$, $P = [\beta_{1,0}, \ldots, \beta_{k,0}]'$ and $W(r') = I_T + r'(d_1A_1 + \ldots + d_kA_k)$, where $x_j = (x_{j1}, \ldots, x_{jT})'$, $A_j = x_{j\circ}C_TC'_{T\circ}x_j$, and the operator denotes the Hadamard product. The LBI test statistic for testing the null hypothesis of deterministic seasonality (H_0: $d_1 + \ldots + d_k = k$) against the alternative hypothesis of mixed deterministic-stochastic seasonality (H_1: $d_1 + \ldots + d_k = r < k$), say $SD_{m,k,n}(r)$, is given by (2) with $K = (d_1A_1 + \ldots + d_kA_k)/k$, which coincides with $RW_{m,k,n}$ when r = k. Noting that $MA_jMA_i = 0$ for $j \neq i$, we find that the eigenvalues of MK have mean and mean-square are given by

$$\bar{\lambda}_1 = (n+1)r/(6k)$$

And

$$\bar{\lambda}_2 = \bar{\lambda}_1(2n^2+7)/30$$

Analogously, when the explanatory variables in (10) are trigonometric seasonal variables ($x_{1t} = 1$, $x_{jt} = \cos(jt\pi/k)$ for j even, and $x_{jt} = \sin[(j-1)t\pi/k]$ for j odd and j > 1), it is convenient to assume that $v_{jt} \sim N(0, r_jS)$, where $r_j = a_jr/k^2$ with $a_j = 1$ (j = 1, k) and $a_j = 2$ (j = 2, ..., k – 1). Now, as before, (10) reduces to (4) when r = k. Here, we can focus our attention on testing the deterministic or stochastic nature of the local level β_{1t}, the (j/2)-th harmonic $\beta_{jt}\cos(\pi jt/k) + \beta_{j+1,t}\sin(\pi jt/k)$ (j = 2, 4, ..., k) or any combination of these k/2 harmonics. Taking as illustration the (j/2)-th harmonic, the LBI test statistic for testing the null hypothesis of deterministic seasonality (H_0: $d_1 + \ldots + d_k = k$) against the alternative of mixed deterministic-stochastic seasonality (H_1: $d_1 + \ldots + d_k = 2$) is given by (2) with $K = a_j(A_j + A_{j+1})/k^2$ with A_j as defined before. We find that the eigenvalues of MK have mean and mean-square given by

$$\bar{\lambda}_1 = a_j(n+1)/(6k)$$

and

$$\bar{\lambda}_2 = \bar{\lambda}_1\left[\frac{(4n^2-1)k^2+15}{60k^2} + b_j\frac{3}{4k^2\sin^2(j\pi/k)}\right]$$

where $b_j = 0$ (j = 1, k) and $b_j = 1$ (j = 2, ..., k – 1). Note that these expressions are also valid to compute the mean and variance of the LBI test statistic for a deterministic level in presence of deterministic seasonality, H_0: $r_1 = 0$

versus H_1: $r_1 > 0$, which is closely related to the KPSS test with seasonal dummies proposed by [23]. Furthermore, [7,8] derived the LBI test statistic, say $TV_{1,k,n}$, for the dual testing problem of deterministic seasonality in presence of a deterministic level, which was generalized to the multivariate case by [16]. This test statistic is given by (2) with $K = (a_2A_2 + ... + a_kA_k)/k^2$. We find that the eigenvalues of MK have mean and mean-square given by

$$\bar{\lambda}_1 = \frac{(n+1)(k-1)}{6k}$$

And

$$\bar{\lambda}_2 = \bar{\lambda}_1\left[\frac{2n^2+7}{30} - \frac{k+1}{4k^2}\right]$$

APPROXIMATE DISTRIBUTIONS AND ACCURACY

Asymptotic Samples

Following [11] and [21], it can be shown that the limiting distribution of $RW_{m,k,n}/mn$ under testing H_0: $r = 0$ is

$$\mathbf{RW_{m,k,n}} \rightarrow \frac{1}{mk}\sum_{j=1}^{\infty}(\pi j)^{-2}\zeta_j \tag{11}$$

where $\xi_j \sim$ iid x^2_{mk}. [2] noted that $RW_{1,1,n}/n$ follows the same limiting distribution as the Cramèr-von Mises goodness-of-fit test statistic. Hence, (11) is the average of mk copies of the Cramèr-von Mises distribution, denoted by CvM(mk)/mk. [24] found that the first four cumulants of a CvM(r) distribution are given by k_1=r/6, k_2=r/45 ,k_3=8r/945,k_4=8r/1575, which reveal that the distribution is strongly rightskewed and leptokurtic. We observe that the two parameter

Inverse Gaussian distribution, IG(μ, λ), can be fitted to possess similar characteristics. The first four cumulants of this distribution are ,

$k_1 = \mu, k_2 = \mu^3 / \lambda, k_3 = 3\mu^5 / \lambda^2, k_4 = 15\mu^7 / \lambda^3$ and matching the first two cumulants, $\mu = r/6$ and $\mu^3/\lambda = r/45$, we obtain that the fitted IG(r/6, $45r^2/6^3$) distribution has third and fourth cumulants given by $\kappa_3 = 8r/900$ and $\kappa_4 = 8r/1350$, which seem to be quite close to those of the CvM (r) distribution. The accuracy of the IG approximation is illustrated in **Figure 1**, which shows the limit pdf of $RW_{1,1,n}/n$ evaluated by the [13] procedure (solid line), along with the pdf of the fitted IG(μ, λ) distribution (dashed line) given by

$$f(x;\mu,\lambda) = [\lambda / (2\pi x^3)]^{1/2} \exp\ [-\lambda(x-\mu)^2 / (2\mu^2 x)]$$

with $x > 0$, $\mu = 1/6$ and $\lambda = 45m/6^3$. We can see that the IG approximation provides a very good fit on both tails of the distribution. Therefore, accurate asymptotic pvalues for $RW_{m,k,n}/mn$ can be computed from the cdf of the IG(μ, λ) distribution.

Finite Samples

The goodness-of-fit in the asymptotic case take us to ask if the approximation will be also good in finite samples. To evaluate the exact null distribution of $RW_{m,k,n}/mn$ from (3)-(6), we must determine the eigenvalues of the matrix MK. To this end, it is convenient to note that the projection matrix M can be alternatively written as M = [Ñ¢$_n$(Ñ$_n$Ñ¢$_n$)$^{-1}$Ñ$_n$] Ä I$_k$. Hence, MK = (Ñ$_n$Ñ¢$_n$)$^{-1}$ Ä I$_k$and its eigenvalues are the reciprocals of those of the tridiagonal matrix Ñ$_n$Ñ¢$_n$, $\lambda_t = [4\sin^2(t\pi/2n)]^{-1}$ (t = 1, 2, ..., n – 1), each one with multiplicity k. In the case m = 1, (3)-(6) can be expressed as a ratio of quadratic forms in normal variables whose distribution was tabulated by [11] using the [13] procedure (they used n – 1 as correction factor instead of n). However, when m > 1, similar tables can be obtained by Monte Carlo simulation from (2)-(6).

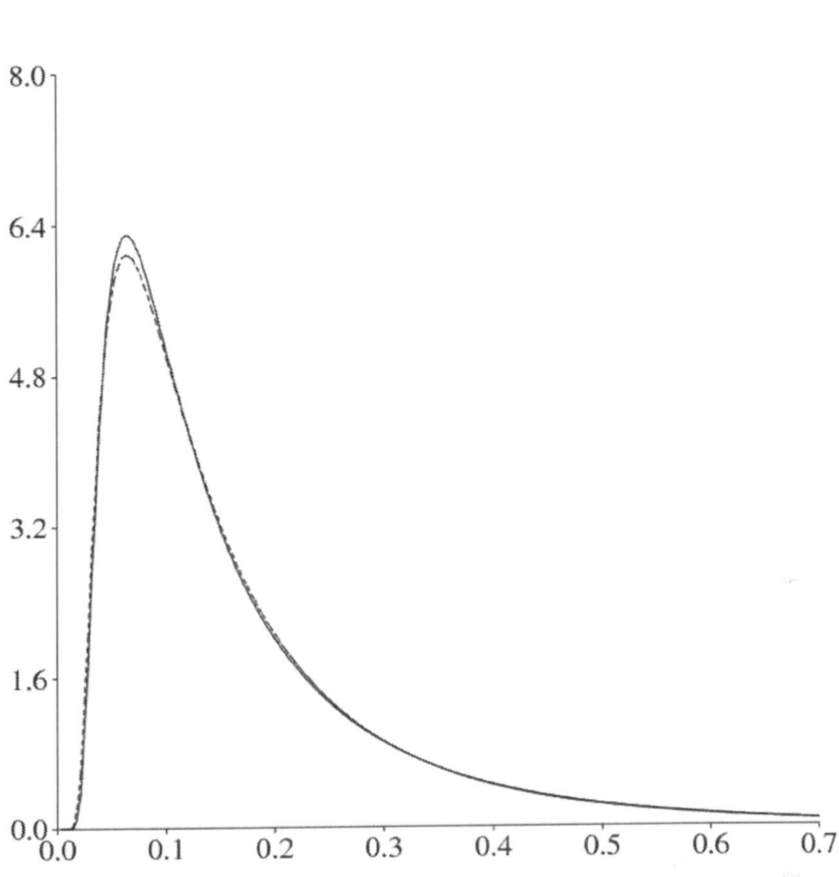

Figure 1: Accuracy of the IG approximation.

To assess the accuracy of the IG approximation in small samples we simply compare the approximate pvalues with the nominal sizes for n = 10, 20, 30, 50 10, k = 1, 2, 3, and m = 1, 2, 3, 4, 5. In general, the approximate p-values agree closely with the nominal sizes even in small samples, being the mean absolute errors less than 0.003. Such discrepancies seem not to be relevant in practical applications. Similar results have been found for $DRW_{m,k,n}$ and $TV_{m,k,n}$. The results of this simulation study are not presented here due to space restrictions but are available from authors at request.

CONCLUSIONS

We have presented seasonal extensions of the [21] test for a deterministic level in multivariate models that can be used to detect different forms of non-invertibility in VARIMA models. The two-moment IG approximation to the null distribution of these test statistics, along with the closed forms expressions for the first two moments here derived, provide a simple and fast way to compute accurate critical values and p-values in practical applications. The proposed approximation could be also useful when modifying the test statistics to deal with intervention variables and serially correlated errors. Finally, the testing procedures described have been implemented in a computer program that can be freely obtained from authors at request.

REFERENCES

1. M. L. King and G. H. Hillier, "Locally Best Invariant Tests of the Error Covariance Matrix of the Linear Regression Model," Journal of the Royal Statistical Society, Series B (Methodological), Vol. 47, No. 1, 1985, pp. 98- 102.
2. J. Nyblom and T. Mäkeläinen, "Comparisons of Tests for the Presence of Random Walk Coefficients in a Simple Linear Model," Journal of the American Statistical Association, Vol. 78, No. 384, 1983, pp. 856-864. doi:10.2307/2288196
3. J. Nyblom, "Testing for Deterministic Linear Trend in Time Series," Journal of the American Statistical Association, Vol. 81, No. 394, 1986, pp. 545-549. doi:10.2307/2289247
4. D. Kwiatkowski, P. C. B. Phillips, P. Schmidt and Y. Shin, "Testing the Null Hypothesis of Stationarity against the Alternative of a Unit Root," Journal of Econometrics, Vol. 54, No. 1-3, 1992, pp. 159-178. doi:10.1016/0304-4076(92)90104-Y
5. J. Nyblom and A. Harvey. "Testing against Smooth Stochastic Trends," Journal of Applied Econometrics, Vol. 16, 2001, pp. 415-429. doi:10.1002/jae.604
6. F. Canova and B. E. Hansen, "Are Seasonal Patterns Constant over Time? A Test for Seasonal Stability," Journal of Business and Economic Statistics, Vol. 13, No. 3, 1995, pp. 237-252. doi:10.2307/1392184
7. M. Caner, "A Locally Optimal Seasonal Unit-Root Test," Journal of Business and Economic Statistics, Vol. 16, No. 3, 1998, pp. 349-356. doi:10.2307/1392511
8. F. Busetti and A. Harvey, "Seasonality Tests," Journal of Business and Economic Statistics, Vol. 21, No. 3, 2003, pp. 420-436. doi:10.1198/073500103288619061
9. K. Tanaka, "Testing for a Moving Average Unit Root," Econometric Theory, Vol. 6, No. 4, 1990, pp. 433-444. doi:10.1017/S0266466600005442

10. P. Saikkonen and R. Luukkonen, "Testing for a Moving Copyright © 2011 SciRes. OJS J. L. GALLEGO, C. DIAZ Copyright © 2011 SciRes. OJS 150 Average Unit Root in Autoregressive Integrated Moving Average Models," Journal of the American Statistical Association, Vol. 88, No. 422, 1993, pp. 596-601. doi:10.2307/2290341
11. W. Tam and G. C. Reinsel, "Tests for Seasonal Moving Average Unit Root in ARIMA Models," Journal of the American Statistical Association, Vol. 92, No. 438, 1997, pp. 725-738. doi:10.2307/2965721
12. W. Tam and G. C. Reinsel, "Seasonal Moving-Average Unit Root tests in the Presence of a Linear Trend," Journal of Time Series Analysis, Vol. 19, No. 5, 1998, pp. 609-625. doi:10.1111/1467-9892.00112
13. J. P. Imhof, "Computing the Distribution of Quadratic Forms in Normal Variables," Biometrika, Vol. 48, No. 3-4, 1961, pp. 419-426. doi:10.1093/biomet/48.3-4.419
14. R. B. Davies, "Numerical Inversion of a Characteristic Function," Biometrika, Vol. 60, No. 2, 1973, pp. 415-417. doi:10.1093/biomet/60.2.415
15. T. W. Anderson and D. A. Darling, "Asymptotic Theory of Certain 'Goodness of Fit' Criteria Based on Stochastic Processes," The Annals of Mathematical Statistics, Vol. 23, No. 2, 1952, pp. 193-212. doi:10.1214/aoms/1177729437
16. F. Busetti, "Tests of Seasonal Integration and Cointegration in Multivariate Unobserved Component Models," Journal of Applied Econometrics, Vol. 21, 2006, pp. 419- 438. doi:10.1002/jae.852
17. J. Nyblom, "Invariant Tests for Covariance Structures in Multivariate Linear Model," Journal of Multivariate Analysis, Vol. 76, 2001, pp. 294-315. doi:10.1006/jmva.2000.1918
18. J. MacKinnon, "Approximate Asymptotic Distribution Functions for Unit-Roots and Cointegration Tests," Journal of Business and Economic Statistics, Vol. 12, 1994, pp. 167-176. doi:10.2307/1391481
19. J. A. Doornik, "Approximation to the Asymptotic Distributions of Cointegration Tests," Journal of Economic Surveys, Vol. 12, 1998, pp. 573-593. doi:10.1111/1467-6419.00068
20. C. R. Rao, "Linear Statistical Inference and its Applications," 2nd Edition, Wiley, New York, 1973. doi:10.1002/9780470316436
21. J. Nyblom and A. Harvey, "Tests of Common Stochastic Trends," Econometric Theory, Vol. 16, 2000, pp. 176-199. doi:10.1017/S0266466600162024
22. A. M. R. Taylor, "Locally Optimal Tests against Unit Roots in Seasonal Time Series Processes," Journal of Time Series Analysis, Vol. 24, No. 5, 2003, pp. 591-612. doi:10.1111/1467-9892.00324
23. P. C. B. Phillips and S. Jin, "The KPSS Test with Seasonal Dummies," Economics Letters, Vol. 77, 2002, pp. 239-243. doi:10.1016/S0165-1765(02)00127-1
24. B. M. Brown, "Cramèr-Von Mises Distributions and Permutation Tests," Biometrika, Vol. 69, No. 3, 1982, pp. 619-624.

CITATION

J. Gallego and C. Díaz, "Testing for Deterministic Components in Vector Seasonal Time Series," Open Journal of Statistics, Vol. 1 No. 3, 2011, pp. 145-150. doi: 10.4236/ojs.2011.13017.

Uncertainty In A Measurement Of Density Dependence On Population Fluctuations

Hiro-Sato Niwa
National Research Institute of Fisheries Science, Yokohama, Japan

6

ABSTRACT

This article discusses the question of how elasticity of the system is intertwined with external stochastic disturbances. The speed at which a displaced system returns to its equilibrium is a measure of density dependence in population dynamics. Population dynamics in random environments, linearized around the equilibrium point, can be represented by a Langevin equation, where populations fluctuate under locally stable (not periodic or chaotic) dynamics. I consider a Langevin model in discrete time, driven by time-correlated random forces, and examine uncertainty in locating the population equilibrium. There exists a time scale such that for times shorter than this scale the dynamics can be approximately described by a random walk; it is difficult to know whether the system is heading toward the equilibrium point. Density dependence is a concept that emerges from a proper coarse-graining procedure applied for time-series analysis of population data. The analysis is illustrated using time-series data from fisheries in the North Atlantic, where fish populations are buffeted by stochastic harvesting in a random environment.

INTRODUCTION

The nature of the negative feedback relationship between population growth rate and abundance is at the heart of population ecology. That said, statistical detection of density dependence using ecological time-series can be problematic. When plotting data on the form of the dependence of population growth rate on abundance, ecologists have been confounded by considerable noise around each relationship [1]. Hassell et al. [2] first pointed out that density-dependent effects are more marked in populations monitored longer. Brook and Bradshaw [3], examining the time-series data of 1198 species, demonstrated a relationship between length of monitoring and increasing evidence for density dependence. However, the duration of a time-series necessary to distinguish a regulated population trajectory from a random walk is still uncertain [4] [5].

A relevant serious concern is the issue of noise color: how does the presence of serial correlation in the external stochastic forcing affect the density dependence in a population? Solow [6] demonstrated that, when a statistical test of density dependence being applied to time series generated by a correlated random walk, the proportion of simulations for which the null hypothesis (random walk model) was rejected was far greater than the significance level for negatively correlated variations (blue noise), and much smaller for positive autocorrelation (red noise). The effects of colored environmental variations on modulating the population elasticity have not been analyzed quantitatively.

This article investigates the analysis of ecological time-series when the aim is to measure density dependence in the demographic processes. The equilibration time [7] [8] of the stochastic process is an informative summary measure of the dynamics of a system, describing a key property of population dynamics [9] [10]. The reciprocal of equilibration time approximates the speed of return to the equilibrium point, i.e. the elasticity (expressing the same concept as the term "density dependence" or "negative feedback" in population dynamics). The objective of this article is to explore the relation between the length of monitoring and the uncertainty in measuring density dependence: how long of a time-series is required to be able to discriminate between density-dependent signal and external noise in the ecological systems? I start with a linear approximation of a discrete-time population dynamics model [11] -[16] , where populations fluctuate slowly under locally stable (not periodic or chaotic) dynamics with low to moderate elasticity [17]. The analysis is illustrated using time-series of annual census of exploited fish populations in the North Atlantic.

POPULATION DYNAMICS

Discrete-Time Langevin Equation

The population renewal process S(t) of marine exploited fishes, responding to stochastic harvesting in a stochastic environment, is described by a discrete-time dynamics

$$S(t+1)=\Lambda\left(S(t)\right)S(t)+\Xi(t+1)-Y(t+1) \tag{1}$$

with a growth-survival factor Λ dependent on spawner abundances S, i.e. the escapement of adults from the harvest Y. The population biomass grows as a result of, not only growth of the surviving escapees in body weight, but also of recruitment of off spring Ξ. During year t an amount Y(t) is harvested with a fraction

$1\text{-}e^{-F(t)}$ to the harvestable adult population, where F(t)=ln[1+Y(t)/S(t)] denotes the fishing mortality.

All density dependence is assumed to be exerted by the adult population S [16]. It is a common observation that recruitment is unrelated to egg numbers over a wide range of spawner abundances [18] -[20]. Stochasticity enters the population dynamics in two ways, both as environmental variation mirrored by recruitment variability and as variation in the fishing mortality. Recruitment is presumed to track auto correlated fluctuations in environmental variables. Fishers would expect this year to get as much catch as last year; there would also be some kind of inertia of fisheries management. There exist memory effects:

adjacent values in the time-series of recruitment and catches-to-escapement ratio are correlated [21].

In a stationary state (with equilibrium quantities denoted by the "$\star$" subscript), Equation (1), log-transformed, gives $\ln(1-w)+F_\star = \ln\wedge(S_\star)$ with the stationary fishing mortality $F_\star = \ln(1+Y_\star/S_\star)$, where $w = \Xi_\star/(S_\star+Y_\star)$ is the stationary ratio of each year's recruitment to the harvestable adult population. The relative deviations from equilibrium point are denoted as $n(t)=\ln(S(t)/S_\star$, $n_1(t) = w\ln(\Xi(t)/\Xi\star)$, and $n_2(t) = F_\star - F(t)$. Natural and human-caused forces (n_1+n_2) are considered as external disturbances. On the assumption that we are free to change Ξ and F independently of S around the equilibrium point, linearizing (log-transformed) Equation (1) yields a difference equation

$$n(t+1)-n(t) = -\gamma n(t)+\sum_{i=1}^{2}\eta_i(t+1) \tag{2}$$

with constant coefficient $\gamma = 1-(1-w)(1+d\ln\wedge/d\ln S)_\star$ evaluated at equilibrium. The coefficient γ measures the (linear) elasticity of the system with respect to change in population size at equilibrium. The successive difference r(t)=n(t+1)-n(t) approximates the annual rate of increase of a population. For generating temporally structured noise, n_i's are both assumed to be described as a first-order autoregressive, AR (1), process with serial correlation coefficients $\lambda \in (-1,1)$ and driven by mean-zero iid shocks. The linear stochastic difference Equation (2) is a discrete-time analogue of the Langevin equation driven by AR(1) processes. The autoregression coefficients are calculated by using the von Neumann's ratio [22] [23], the ratio of the mean square successive difference $\sigma^2_{\Delta n_i}$ to the variance $\sigma^2_{n_i}$ for AR(1) variable n_i, that is, $\lambda_i = 1-\sigma^2_{n_i}/2\sigma^2_{n_i}$. Equation (2) is iterated to give the Δt-year difference of population size

$$n(t+\Delta t)-n(t) = -\left(1-\lambda_0^{\Delta t}\right)n(t)+\sum_{i=1}^{2}\sum_{j=0}^{\Delta t-1}\lambda_0^{j}\eta_i(t+\Delta t-j) \tag{3}$$

With $\lambda_0 = 1-\gamma$.

Equilibration Time

The population process with multiple decay-rate constants $(-\ln \lambda_i, i \in \{0,1,2\})$ is considered to calculate the asymptotic decay-time constant T_{eq}, i.e. the total equilibration time; the approach follows Roughgarden [11]. Let $\hat{\sigma}^2_{\eta_i} = \sigma^2_{\eta_i} / (1-\lambda_0\lambda_i)$. Taking the mean squares over Equations (2) and (3) mediates the following two relations, respectively,

$$\sigma_r^2 / 2\sigma_n^2 = \gamma - \sum_{i=1}^{2} \hat{\sigma}^2_{\eta_i} \lambda_i / \sigma_n^2 \tag{4}$$

with the ratio of the mean square successive difference σ_r^2 to the variance σ_n^2 in population size n(t), and

$$\gamma \sum_{\Delta t=0}^{\infty} \rho(\Delta t) = 1 + \sigma_n^{-2} \sum_{i=1}^{2} \hat{\sigma}^2_{\eta_i} \lambda_i / (1-\lambda_i) \tag{5}$$

with the autocorrelation function of series n(t) at lag Δt years

$$\rho(\Delta t) = \lambda_0^{\Delta t} + \sigma_n^{-2} \sum_{i=1}^{2} \hat{\sigma}^2_{\eta_i} \lambda_i \frac{\lambda_0^{\Delta t} - \lambda_i^{\Delta t}}{\lambda_0 - \lambda_i}, \tag{6}$$

where the correction (the second term on the right-hand-side) is due to memory effects of external perturbations. From the recursion, $n(t) = \Sigma_{i=1}^{2} \Sigma_{j=0}^{\infty} \lambda_0^j \eta_i (t-j)$, a corollary to Equation (5) is obtained:

$$\sigma_n^2 = \sum_{i=1}^{2} (1+\lambda_0\lambda_i) \hat{\sigma}^2_{\eta_i} / \left(1-\lambda_0^2\right). \tag{7}$$

Equations (4) and (7) yield the quadratic equation for the elasticity γ

$$\gamma^2\sigma_n^2 + (1-\gamma)\sigma_r^2 = \sum_{i=1}^{2}\sigma_{\eta_i}^2 . \tag{8}$$

Thus, the (linear) response of a system to external perturbations is expressed in terms of fluctuation properties of the system in equilibrium. The elasticity and the variance of population fluctuations cannot be independent, but they are related to each other in the equilibrium system [24].

The average relaxation time of population fluctuations defines the total equilibration time T_{eq} as follows [24]

$$\sum_{\Delta t=0}^{\infty}\rho(\Delta t) = \sum_{\Delta t=0}^{\infty}\exp(-\Delta t/T_{eq}). \tag{9}$$

After the time T_{eq} memory of the initial conditions is lost: the deviation from the equilibrium is expected to decay away in an exponential fashion with a time constant given by the equilibration time T_{eq}. The total elasticity with respect to change in population size measures the strength D of total density dependence [10] [16]:

$$\mathrm{E}\left[n(t+1)-n(t)\middle|n(t)=n\right] = -Dn,$$

i.e. the expectation of change in population size given n(t)=n is equal to 100 × D percent negative feedback. Since the total density dependence is the asymptotic multiplicative growth rate of population per year, -ln(1-D) reads the asymptotic decay-rate constant. Smaller values of D indicate that systems return to equilibrium slower. Since the exponential-decay time is the inverse of the decay-rate constant, the total density dependence reads

$$D = 1 - e^{-1/T_{eq}}.$$

Therefore, the density dependence D and the variance of population fluctuations are related to each other through the relation $D^{-1} = \Sigma_{\Delta t=0}^{\infty} \rho(\Delta t)$ with Equation (5).

Indeterminacy in Population Dynamics

Equation (2) with (8) yields

$$\frac{n(t+1)-n(t)}{\sigma_r} = -\gamma \frac{\sigma_n}{\sigma_r} \frac{n(t)}{\sigma_n} + \sqrt{1-\gamma+\gamma^2 \frac{\sigma_n^2}{\sigma_r^2}} \frac{\sum_{i=1}^{2} \eta_i (t+1)}{\sqrt{\sum_{i=1}^{2} \sigma_{\eta_i}^2}}$$

(where each term is standardized), implying that, if a population is governed by slowly damped dynamics $(0 < \gamma < 1)$, external noises mask signals for evidence of deterministic behavior of the system. Unless the population exhibits damped-oscillator dynamics (i.e. over-compensation, $(1 < \gamma < 2)$, the relationship between r(t) and n(t) is characterized by large variance in growth rate; most of the changes in population size occur in a density-independent manner. If a perturbation brings the system away from the equilibrium point such that

$|n| > \sigma_n \sqrt{1 + (1-\gamma) / (\gamma^2 \sigma_n^2 / \sigma_r^2)}$, then the deterministic signal becomes visible over the noise; using Equations (4) and (5) this inequality reduces to

$$|n| \gtrsim \sigma_n \sqrt{L_c / \pi}$$

with

$$L_c = \left[1 + 2 \Big/ \left(e^{1/T_{eq}} - 1\right)\right] \pi.$$

When one performs some observations for L years, the uncertainty in locating the population equilibrium, $\Delta\bar{n} = \mathrm{SD}\left[\bar{n}\right]$, i.e. the standard deviation of the sample mean $\bar{n} = \Sigma_{t=1}^{L} n(t)/L$, propagates into the density dependence. The variance of $\bar{n}$ is written as [13] [16], $\mathrm{Var}\left[\bar{n}\right] = \sigma_n^2 \Sigma_{i,j=1}^{L} \rho(i-j)/L^2$. Making use of the explicit form for the autocorrelation function (6) yields

$$\mathrm{Var}\left[\bar{n}\right] = \frac{\sigma_n^2}{L}\left[2\sum_{\Delta t=0}^{\infty} \rho(\Delta t) - 1 + \frac{\varsigma}{L}\right]$$

with

$$\varsigma = -\frac{2\lambda_0\left(1-\lambda_0^L\right)}{\left(1-\lambda_0\right)^2} - \sum_{i=1}^{2} \frac{\hat{\sigma}_{\eta_i}^2}{\sigma_n^2} \frac{2\lambda_i}{\lambda_0 - \lambda_i}\left[\frac{\lambda_0\left(1-\lambda_0^L\right)}{\left(1-\lambda_0\right)^2} - \frac{\lambda_i\left(1-\lambda_i^L\right)}{\left(1-\lambda_i\right)^2}\right].$$

By virtue of Equation (9), a time-equilibrium uncertainty relation is obtained:

$$\Delta\bar{n}/\sigma_n = \sqrt{L_c/L\pi} + \mathcal{O}\left(L^{-3/2}\right), \tag{10}$$

Where $\mathcal{O}(L^{-3/2})$ denotes a remainder term of order $L^{-3/2}$ given by $\varsigma L^{-3/2}\sqrt{\pi/4L_c}$. Thus, L_c provides another measure of the time scale of population dynamics, and reads $L_c = 2\pi T_{eq} + O(T_{eq}^{-1})$ for D<1 with large valued

T_{eq}. Any observation over a short duration is associated with large indeterminacy. The supposed population equilibrium $\ln S_*$ (chosen as the sample mean) and the length of time-series data are complementary: the uncertainty of population equilibrium is decreased by increasing the duration of observation. L_c is designated the complementary time. It is worth pointing out that the complementary time is always positive for 0<D<2, while the equilibration time can acquire either non-negative real number or complex number.

Coarse Graining

Let C_L be the degree of certainty, with which the supposed equilibrium lies within the bounds $E\left[n(t)\middle|n(t) >< 0\right]$:

$$C_L = \Pr\left(\left|\overline{n}\right| \leq \sigma_n \sqrt{2/\pi}\right)$$

for normally distributed population fluctuations (with the true (population) standard deviation σ_n). This is equivalent to $\Pr\left(\left|\overline{n}\right|\right) < \Delta n \sqrt{2L/L_c}$. Let $\overline{[\bullet]}_{\pm} \left(= \Sigma_{\{t|n(t)><0\}} [\bullet] / \#_n(t) >< 0\right)$ be the sample mean, taken under the condition that n is larger (smaller) than the supposed equilibrium point, where $\#_{n(t)><0}$ is the number of observations $n(t) >< 0$ in the time-series. The negative relationship of conditional sample mean of growth rate, $\overline{[r]}_{\pm}$, versus $\overline{[n]}_{\pm}$ is expected to be visible with a probability C_L. The larger the probability C_L, the more accurately the density dependence D is measured from the population time-series.

In order to calculate the probability C_L, I analyze the time-series at different resolutions by constructing a coarse-grained time-series. Let $\lceil \cdot \rceil$ and $\lfloor \cdot \rfloor$ be the ceil and the floor functions. The coarse-grained time-series is built by taking the average inside a non-overlapping moving window with $\lceil T_{eq} \rceil$ data points,

$$n_{T_{eq}}(j) = \sum_{i=(j-1)\lceil T_{eq} \rceil + 1}^{j\lceil T_{eq} \rceil} n(i) / \lceil T_{eq} \rceil, \quad j = 1, \cdots, \left\lfloor L / \lceil T_{eq} \rceil \right\rfloor.$$

The coarse-grained time-series are considered to be serially independent. Accordingly, the uncertainty of population equilibrium (the standard deviation of the mean) is calculated to be

$$\Delta \overline{n} = \frac{\mathrm{SD}\left[n_{T_{\mathrm{eq}}} \right]}{\left\lfloor L / \left\lceil T_{\mathrm{eq}} \right\rceil \right\rfloor^{1/2}},$$

and the statistic

$$\frac{\overline{n}}{\Delta \overline{n}} = \frac{\overline{n}_{T_{\mathrm{eq}}}}{\mathrm{SD}\left[n_{T_{\mathrm{eq}}} \right] / \left\lfloor L / \left\lceil T_{\mathrm{eq}} \right\rceil \right\rfloor^{1/2}} \tag{11}$$

has a standard Gaussian distribution. In reality, we only have the sample standard deviation. So, replacing $\mathrm{SD}\left[n_{T_{eq}} \right]$ in Equation (11) by the square root of the sample variance, $\left(\Sigma_{j=1}^{\lfloor L/\lceil T_{eq} \rceil \rfloor} n_{T_{eq}}^2 (j) / \left(\lfloor L / \lceil T_{eq} \rceil \rfloor - 1 \right) \right)^{1/2}$, yields a statistic which has Student's t-distribution with $\lfloor L / \lceil T_{eq} \rceil \rfloor - 1$ degrees of freedom (d.f.); we can use the Student's t-statistic to see how good the estimate of equilibrium population size. The value of C_L, i.e. the integral of Student's t-PDF between $\pm\sqrt{2L / L_c}$, quantifies the ability to infer density dependence from a given (long) observation time-series, and $\sqrt{2L / L_c}$ is the two-tailed $100 \times (1 - C_L)$ percentage point of Student's t-distribution with d.f.= $\lfloor L / \lceil T_{eq} \rceil \rfloor - 1$. Accordingly, the following holds

$$\sqrt{2L/L_c} = t_{(1-C_L)/2} \left(\left\lfloor L / \left\lceil T_{\mathrm{eq}} \right\rceil \right\rfloor - 1 \right) \tag{12}$$

(precisely the left-hand-side involves the remainder term of order $L^{-1/2}$). After L years' observation the negative relationship is visible with a probability of C_L. Equation (12) with $L=L_c$ and $L_c = 2\pi T_{eq}$ yields

$C_L\left[=\Pr\left(\left|\bar{n}/\Delta\bar{n}\right|\right)\leq\sqrt{2}\right]=0.784$, implying that, when the observation series is longer than L_c years, the negative relationship is visible with a probability of 78.4% or more. When monitoring an equilibrium population (in the dynamic balance) for L years, the probability of the true equilibrium lying outside the bounds $\left(\pm\sigma_n\sqrt{2/\pi}\right)$ reads $1-C_L$; the sample mean growth rate $\overline{[r]}_{\pm} >< 0$ will then be observed, even when $L>L_C$. The density dependence will be invisible with a probability of $1-C_L$, that is, it has a tendency to overshoot and to leave the center point $\bar{n}$ (i.e. sample average).

North Atlantic Fisheries

Empirical analysis [21] [25] suggests that population time-series from the North Atlantic fisheries are asymptotically stationary. The notion of asymptotic stationarity means that the probability density function of the variable monitored over a wide time interval exists and it is uniquely defined. The logarithmic changes in annual spawner abundance,r(t), are Gaussian distributed and the observed mean r's are not significantly different from zero. There is no important overall trend in spawner abundance, as expected in the equilibrium system [3].

I now apply the theory for measuring density dependence to time series of North Atlantic commercial species, which are the same as the fish stocks analyzed in [26] ; there are 38 populations of ten species, cod (Gadus morhua), haddock (Melanogrammus aeglefinus), herring (Clupea harengus), mackerel (Scomber scombrus), plaice (Pleuronectes platessa), saithe (Pollachius virens), sardine (Sardina pilchardus), sole (Solea solea), sprat (Sprattus sprattus), and blue whiting (Micromesistius poutassou). Data are extracted from the 2008 working group reports of the International Council for the Exploration of the Sea (ICES) [27].

The length of examined time-series ranges from 23 to 60 years. Estimates of model parameters are plotted against the equilibration time

T_{eq} in Figure 1. The estimates of D and L_C are $0.06 \leq D \leq 0.43$ and $11.5 \leq L_c \leq 93.7$ years.

For a sufficiently long Δt, Equation (3) with medium elasticity $(-1 < \lambda_0 < 1)$ yields

$$n(t+\Delta t) - n(t) = -n(t) + \eta_\infty(t+\Delta t),$$

Where η_∞ denotes time-independent random forcing with mean 0 and variance σ_n^2. This coarse-grained equation implies that, while for $|n| < \sigma_n$ the signals (Δt-year growth rate) are at the level of noise, if the system is displaced $(|n| < \sigma_n)$ away from the equilibrium point, the negative relationship becomes visible over the noise on time scales longer than the marginal, equilibration time. Figure 2 demonstrates that though, when plotting annual growth rate against population size, the relationship well looks like a shotgun pattern, when

$y = n(t + \lceil T_{eq} \rceil) - n(t)$ is plotted against x=n(t), the negative feedback effects are visible, i.e. $y >< 0$ for $x >< \pm\sigma_n$.

After averaged over sizes $n >< 0$ in the time-series data over L>L_c years, the negative relationship can be seen with a predicted probability $C_L \gtrsim 0.78$, whereas the dependence of $\overline{[r]}_\pm$ on $\overline{[n]}_\pm$ in the data comprising

L<L_c years is rather vague (upper panels of Figure 3). Besides, I perform simulations to validate the criterion for visibility of density-dependent relationship. For simulations based on the population equation of motion (2) with the AR(1) model for the variations in the recruitment and the fishing mortality, I use parameter values that are obtained by analyzing the ICES time-series data sets. When plotting the simulated diagrams of mean growth rate versus mean abundance, conditioned on n><0, for the populations observed over L>L_c years, the data are divided into two piles; for L>$L_{c'}$, the gap between the piles is closed up (lower panels

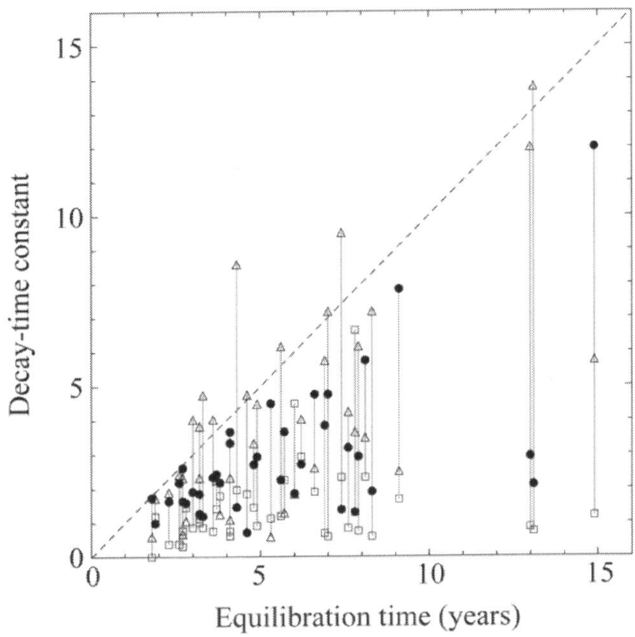

Figure 1: Estimates of decay-time constants of the North Atlantic systems. The equilibration time T_{eq} of each population is compared with the decay-time constants $\left(-1/\ln|\lambda_i|, i \in \{0,1,2\}\right)$ inherent in the system and for the time-series of recruitment and catches-toescapement ratio. The solid circles refer to λ_0, and the open squares (triangles) to recruitment (annual catch) variability. The dashed line represents y = x.

of Figure 3). Of 1000 simulations for each population, the proportion of simulations in which the negative feedback on the sample mean growth rate is observed (i.e. $\overline{[r]}_{\pm} >< 0$) is plotted in Figure 4.

INTERTWINING OF STOCHASTIC PROCESSES

In this section, I discuss the effects of multiple noise sources and noise color on measuring density dependence.

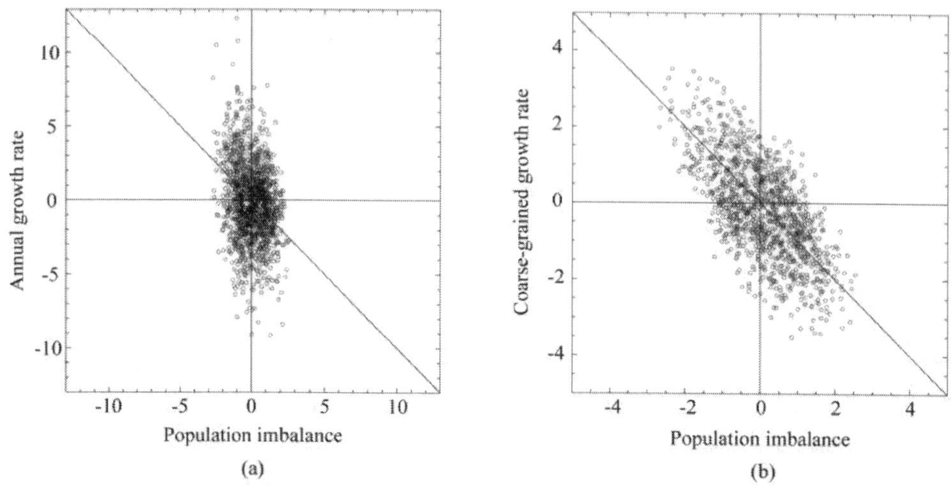

Figure 2: Population growth diagram of the North Atlantic fishes on fineand coarse-grained timescales. (a) The diagrams of annual growth rate (divided by total density dependence D) versus population abundance of the examined fishes are aggregated and plotted. (b) The coarse-grained diagrams demonstrate that the negative relationship is visible between T_{eq}-year growth rate and population imbalance from equilibrium. The axes are scaled by σ_n. The lines represent $y = -x$.

Delayed Density Dependence

Recently, Ives et al. [10] developed a pragmatic approach based on the autoregressive moving average (ARMA) model to assessment of ecological time-series, and identified the lagged structure in the data as the ARMA order. The time-lagged density-dependent effect in population fluctuations appears: the direction of annual change in population size depends on the past population trajectory.

Consider the system perturbed by multiple noise sources, where external perturbations η_i follow AR(1) with autoregression coefficient λ_i, driven by mean-zero iid random shocks ε_i, under the assumption of mutual independence of η_i's $(i = 1,\ldots,p-1$; note that the results obtained in the above section can be generalized to an arbitrary number

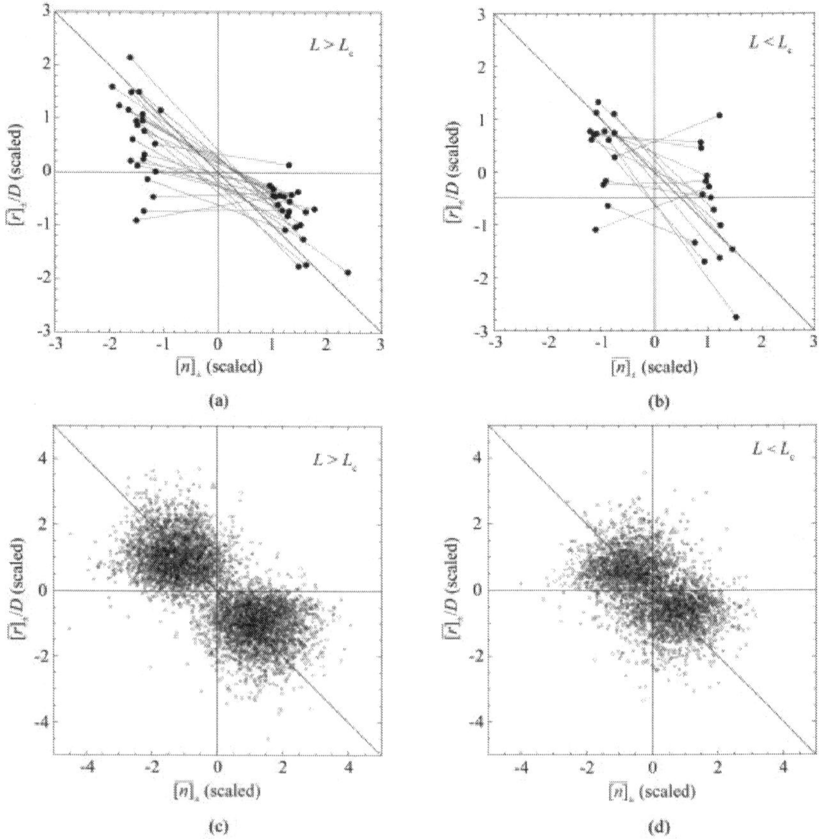

Figure 3: Mean behavior of fish populations in the North Atlantic. Shown are the diagrams (scaled by $\sqrt{2\Delta\bar{n}}$) of mean growth rate (divided by D) versus mean abundance, conditioned on a current population imbalance $n(t)><0$. For the populations observed over $L > L_c$ years (panel (a)), the negative feedback becomes statistically visible with a probability of $C_L \gtrsim 0.78$. For 15 of the 38 populations examined, the time-series data comprise $L < L_c$ years (panel (b)). In panels (c) and (d) (showing the counterparts of panel (a) and (b), respectively), the simulated growth diagrams of North Atlantic fish populations are depicted. 100 pairs for each population are piled up.

of environmental variables). Start by picking a variable η_i to eliminate, and substitute $\eta_i(t) = n(t) - \lambda_0 n(t-1) - \Sigma_{i=2}^{p-1}\eta_i(t)$ into both sides of the AR(1) equation $\eta_i(t+1) = \lambda_1\eta_i(t) + \varepsilon_1(t+1)$; this procedure, repeated to eliminate $\eta_i(i = 2,\ldots,p-1)$ one by one from the simultaneous equations, yields the ARMA $(p,p-2)$ model

$$n(t) = \sum_{\tau=1}^{p} \beta_\tau n(t-\tau) + \sum_{\tau=0}^{p-2} \boldsymbol{\alpha}_\tau \cdot \boldsymbol{\varepsilon}(t-\tau)$$

with (p-1)-dimensional vectors $\alpha_\tau = (\alpha_{\tau,1,},\ldots,\alpha_{\tau,p-1})$ and $\varepsilon(t) = (\varepsilon_1(t),\ldots,$ $\varepsilon_{p-1}(t))$. The ARMA coefficients

β_i and α_{ij} are extracted by expanding the right-hand-side of the following pthand (p-2)nd-degree polynomials

$$\sum_{\tau=0}^{p} \beta_\tau x^{p-\tau} = -\prod_{i=0}^{p-1}(x-\lambda_i), \tag{13}$$

$$\sum_{\tau=0}^{p-2} \alpha_{\tau j} x^{p-2-\tau} = \prod_{i=1(i\neq j)}^{p-1} (x-\lambda_i), \quad j = 1,2,\cdots,p-1.$$

Since the characteristic equation for the pth-order AR component is $\Sigma_{i=0}^{p}\beta_i x^{p-i} = 0$, the roots are found from Equation (13) to be $\lambda_i(i = 0,1,\ldots,p-1)$. We see that the delayed density dependence in population dynamics arises from serial correlation in the external stochastic forcing [12] [24].

Equation (2), transformed into the ARMA(3,1) form, gives

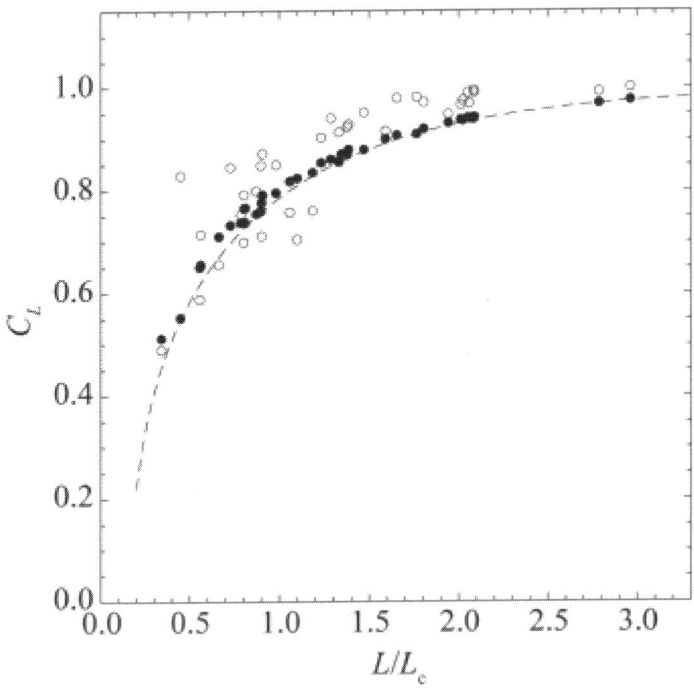

Figure 4: Degree of certainty. The values of C_L are plotted against L/L_c(solid circles for the North Atlantic fish populations). The dashed line is calculated from Equation (12) with the approximation $L_c = 2\pi T_{eq}$. The open circles read C_L estimated from 1000 simulated data sets.

$$\Delta^3 n(t) = -\sum_{i=0}^{2}(1-\lambda_i)\Delta^2 n(t) - \sum_{i<j}(1+\lambda_i\lambda_j)\Delta n(t) - \prod_{i=0}^{2}(1-\lambda_i)n(t) + \sum_{\tau=0}^{1}\alpha_\tau \cdot \varepsilon(t+3-\tau),$$

Where Δ denotes the differencing operation, and $\{\varepsilon(t)\}$ is a sequence of two-dimensional independent random vectors. This third-order stochastic difference equation of population motion does not have an oscillatory solution (except for the period of two years), because λ_i's are restricted between ±1. Figure 1 shows that the time constant of the dominant mode (with the slowest decay rate determined by the maximum eigenvalue) disagrees with the equilibration time T_{eq}. The differential coefficient of T_{eq} with respect to the variance $\sigma^2_{\eta_1}$, for fixed constant λ_1, is readily seen from Equation (5), as

$$\left[\frac{\partial T_{\text{eq}}}{\partial \sigma^2_{\eta_1}}\right]_{\lambda_1=\text{const}} = -(\lambda_2 - \lambda_1)\sigma^2_{\eta_2} Q$$

with

$$Q = \frac{\left(e^{1/T_{\text{eq}}} - 1\right) T^2_{\text{eq}} \left(1 + \prod_{i=0}^{2} \lambda_i\right)}{\sigma^4_n (1-\lambda_0)^2 (1+\lambda_0) \prod_{i=1}^{2} (1-\lambda_i)(1-\lambda_0\lambda_i)},$$

where the variance $\sigma^2_{\eta_1}$ varies linearly with the variance of ε_1. Accordingly, the equilibration time T_{eq} is not determined solely by the eigenvalues λ_i, but depends on the amplitude σ_n of population fluctuations through the driving noise ε_i of the AR(1) process $\eta_i(t)$.

Noise Color

I have shown that apparently density-independent influences, which arise from colored environmental variations, modulate the population elasticity. In the following let us further analyze the impact of color of environmental variation on the total density dependence in a population. Here, for mathematical simplicity, the effects of recruitment fluctuations η_1 are only taken into account on the variance in population abundance, where the inherent elasticity γ of the system is a given constant; the variance $\sigma^2_{\eta_1}$ varies with the value of AR(1) coefficient λ_1 (the variance of ε_1 is fixed at a constant).

Equations (4), (5) and (7) describe the interaction between endogenous dynamics of the population and external disturbances. I demonstrate, in Figure 5, how color of recruitment variation affect the population fluctuations.

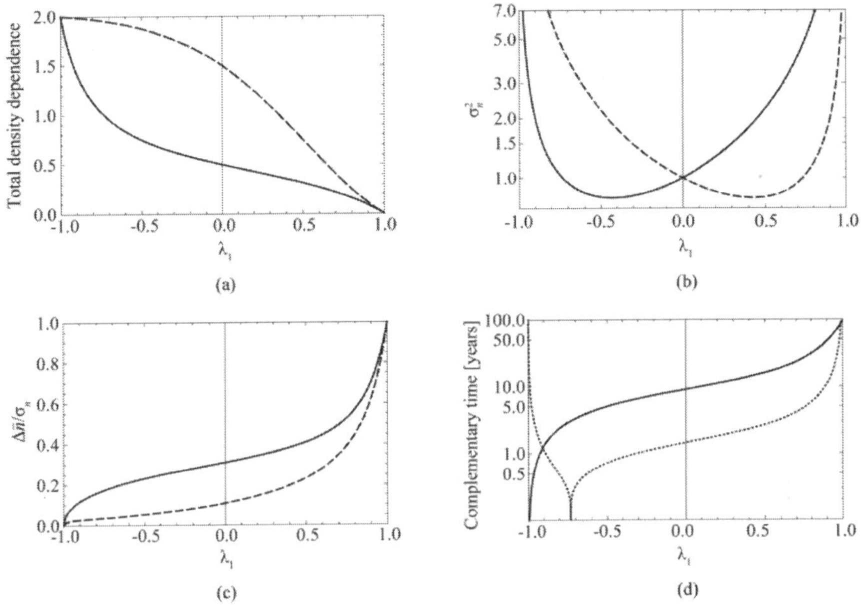

Figure 5. Effects of noise color on population processes. The total density dependence D (panel (a)), the variance σ_n^2 of population fluctuations (panel (b), semi-log scale), the uncertainty $\Delta n / \sigma_n$ of population equilibrium (panel (c)), and the complementary time $\left(L_c + \pi\zeta / L\right)$ including the remainder term (panel (d), semi-log scale) are plotted as a function of environmental autocorrelation λ_1, where L = 30 and the inherent elasticity γ of the system is fixed as $\lambda_0 (= 1 - \gamma) = 0.5$ (solid line) or −0.5 (dashed line). In panel (b), σ_n^2 is divided by the variance of population process driven by white noise. In panel (d), the dotted line represents the equilibration time T_{eq} (indicating $-1/\ln|1-D|$ for D > 1), where $\lambda_0 = 0.5$.

The total density dependence D decreases when redness of noise color is increased (an increase in the autocorrelation is denoted "increased redness"), and $D \to 0$ as $\lambda_1 \to 1$ for any $\gamma \in (0,2)$. The reduction in negative feedback on population growth leads to a slowing down of the equilibration process: population fluctuations have very long relaxation times and the system returns to equilibrium very slowly. In the limit $D \to 0$, the population imbalance follows a random walk irrespective of the values of γ. In a blue environment, $D \to 2$ as $\lambda_1 \to -1$ for any γ. In the limit $D \to 2$, the system is expected to behave as $n(t+1) = -n(t) + \varepsilon(t+1)$ (with a mean-zero iid random variable $\varepsilon(t)$), which is iterated to yield $n(t+1) = -n(t-1) + [\varepsilon(t+1) - \varepsilon(t)]$; consequently, the amplitude jitter in the population oscillation exhibits a random-walk property and is non-stationary (regardless of the value γ). In the limit $\lambda_1 \to -1$, while the system exhibits negative feedback, the population is not regulated. Both $\lambda_1 \to \pm 1$ lead to reduction in population regulation, resulting in unstable dynamics: the amplitude σ_n of population fluctuations diverges asymptotically. Either way, $\lambda_0 (= 1 - \gamma) >< 0$, the variance σ_n^2 reaches a minimum at $\lambda_0 = -\lambda_0 + \mathcal{O}(\lambda_0^3)$. Note from Equation (13) that when $\Sigma_i \lambda_i = 0$, the temporal correlation of population fluctuations significantly decreases (fluctuations become white in time) and the population follows rapidly mean-reverting stochastic process. The effect of an increase in redness of noise is to enhance the uncertainty in locating the population equilibrium and the complementary time. In the limit $\lambda_1 \to 1$, we are far from the allowed certainty in measuring density dependence, $\Delta \bar{n} / \sigma_n \to 1$ for any γ, regardless of L. It is also confirmed that $\Delta \bar{n} \to \infty$ in both the limits $\lambda_1 \to \pm 1$; while in the limit $\lambda_1 \to -1$, the uncertainty of population equilibrium diverges less rapidly than the standard deviation of population fluctuationsi. ie. $\Delta \bar{n} / \sigma_n \to \sqrt{1 - (-1)^L} / \sqrt{2L}$.

CONCLUDING REMARK

Analyzing a stochastic process with slowly damped dynamics, I have derived the expression of complementarity between uncertainty in locating the population equilibrium and length of monitoring. The complementarity relation imposes a limit on the degree of certainty to which a measurement of density dependence is known. The complementarity is an essential criterion for disentangling density-dependent signals from external noise in the population process. Looking at a long time-scale makes the negative feedback on population fluctuations visible. It is difficult to know whether the system is heading toward the equilibrium point in the time-series of length less than the complementary time. This implies that density dependence is a concept that emerges from applying a proper coarse-graining procedure for time-series analysis.

The approach taken in this article is based on linear approximation of a nonlinear stochastic model. A linear approximation applies for "small" perturbations from an equilibrium point, where the "smallness" should be measured relative to the range of the uncertainty of population equilibrium. Because a large uncertainty exists in determining the population equilibrium from short observation series in ecology, it is probable that the system spends most of its time "near" the equilibrium point; a linear approximation would be usually enough accurate.

REFERENCES

1. Sibly, R.M. and Hone, J. (2002) Population Growth Rate and its Determinants: An Overview. Philosophical Transactions of the Royal Society of London B, 357, 1153-1170.http://dx.doi.org/10.1098/rstb.2002.1117
2. Hassell, M.P., Latto, J. and May, R.M. (1989) Seeing the Wood for the Trees: Detecting Density Dependence from Existing Life-Table Studies. Journal of Animal Ecology, 58, 883-892. http://dx.doi.org/10.2307/5130
3. Brook, B.W. and Bradshaw, C.J.A. (2006) Strength of Evidence for Density Dependence in Abundance Time Series of 1198 Species. Ecology, 87, 1445-1451. http://dx.doi.org/10.1890/0012-9658(2006)87[1445:SOEFDD]2.0.CO;2
4. Connell, J.H. and Sousa, W.P. (1983) On the Evidence Needed to Judge Ecological Stability or Persistence. American Naturalist, 121, 789-824.http://dx.doi.org/10.1086/284105
5. Hixon, M.A., Pacala, S.W. and Sandin, S.A. (2002) Population Regulation: Historical Context and Contemporary Challenges of Open vs. Closed Systems. Ecology, 83, 1490-1508. http://dx.doi.org/10.1890/0012-9658(2002)083[1490:PRHCAC]2.0.CO;2
6. Solow, A.R. (1990) Testing for Density Dependence: A Cautionary Note. Oecologia, 83, 47-49. http://dx.doi.org/10.1007/BF00324632
7. Beddington, J.R. and May, R.M. (1977) Harvesting Natural Populations in a Randomly Fluctuating Environment. Science, 197, 463-465.http://dx.doi.org/10.1126/science.197.4302.463
8. Shepherd, J.G. and Horwood, J.W. (1979) The Sensitivity of Exploited Populations to Environmental "Noise", and the Implications for Management. Journal du Conseil International pour l'Exploration de la Mer, 38, 318-323.http://dx.doi.org/10.1093/icesjms/38.3.318

9. Ives, A.R., Dennis, B., Cottingham, K.L. and Carpenter, S.R. (2003) Estimating Community Stability and Ecological Interactions from Time-Series Data. Ecological Monographs, 73, 301-330. http://dx.doi.org/10.1890/0012-9615(2003)073[0301:ECSAEI]2.0.CO;2
10. Ives, A.R., Abbott, K.C. and Ziebarth, N.L. (2010) Analysis of Ecological Time Series with ARMA(p, q) Models. Ecology, 91, 858-871. http://dx.doi.org/10.1890/09-0442.1
11. Roughgarden, J. (1975) A Simple Model for Population Dynamics in Stochastic Environments. American Naturalist, 109, 713-736. http://dx.doi.org/10.1086/283039
12. Royama, T. (1981) Fundamental Concepts and Methodology for the Analysis of Animal Population Dynamics, with Particular Reference to Univoltine Species. Ecological Monographs, 51, 473-493. http://dx.doi.org/10.2307/2937325
13. McArdle, B.H. (1989) Bird Population Densities. Nature, 338, 628.http://dx.doi.org/10.1038/338628a0
14. Lundberg, P., Ranta, E., Ripa, J. and Kaitala, V. (2000) Population Variability in Space and Time. Trends in Ecology & Evolution, 15, 460-464. http://dx.doi.org/10.1016/S0169-5347(00)01981-9
15. Jonzén, N., Ripa, J. and Lundberg, P. (2002) A Theory of Stochastic Harvesting in Stochastic Environments. American Naturalist, 159, 427-437.http://dx.doi.org/10.1086/339456
16. Lande, R., Engen, S., Sæther, B.-E., Filli, F., Matthysen, E. and Weimerskirch, H. (2002) Estimating Density Dependence from Population Time Series Using Demographic Theory and Life-History Data. American Naturalist, 159, 321- 337. http://dx.doi.org/10.1086/338988
17. Edelstein-Keshet, L. (2005) Mathematical Models in Biology. Society for Industrial and Applied Mathematics. http://dx.doi.org/10.1137/1.9780898719147
18. Shepherd, J.G. and Cushing, D.H. (1990) Regulation in Fish Populations: Myth or Mirage? Philosophical Transactions of the Royal Society of London B, 330, 151-164.http://dx.doi.org/10.1098/rstb.1990.0189
19. Walters, C. and Parma, A.M. (1996) Fixed Exploitation Rate Strategies for Coping with Effects of Climate Change. Canadian Journal of Fisheries and Aquatic Sciences, 53, 148-158. http://dx.doi.org/10.1139/f95-151
20. Dixon, P.A., Milicich, M.J. and Sugihara, G. (1999) Episodic Fluctuations in Larval Supply. Science, 283, 1528-1530. http://dx.doi.org/10.1126/science.283.5407.1528
21. Niwa, H.S. (2006) Recruitment Variability in Exploited Aquatic Populations. Aquatic Living Resources, 19, 195-206. http://dx.doi.org/10.1051/alr:2006020
22. von Neumann, J. (1941) Distribution of the Ratio of the Mean Square Successive Difference to the Variance. Annals of Mathematical Statistics, 12, 367-395.http://dx.doi.org/10.1214/aoms/1177731677
23. Bulmer, M.G. (1975) The Statistical Analysis of Density Dependence. Biometrics, 31, 901-911. http://dx.doi.org/10.2307/2529815

24. Kubo, R., Toda, M. and Hashitsume, N. (1991) Statistical Physics II. Nonequilibrium Statistical Mechanics. 2nd Edition, Springer-Verlag, Berlin.http://dx.doi.org/10.1007/978-3-642-58244-8
25. Niwa, H.S. (2007) Random-Walk Dynamics of Exploited Fish Populations. ICES Journal of Marine Science, 64, 496- 502. http://dx.doi.org/10.1093/icesjms/fsm004
26. Sparholt, H., Bertelsen, M. and Lassen, H. (2007) A Meta-Analysis of the Status of ICES Fish Stocks during the Past Half Century. ICES Journal of Marine Science, 64, 707-713.http://dx.doi.org/10.1093/icesjms/fsm038
27. ICES (2008) Report of the ICES Advisory Committee, ICES Advice 2008, Books 1-10.http://www.ices.dk/publications/our-publications/Pages/ICES-Advice.aspx

CITATION

Hiro-Sato Niwa Uncertainty in a Measurement of Density Dependence on Population Fluctuations DOI:10.4236/am.2014.58104

Truncated Geometric Bootstrap Method for Time Series Stationary Process

T. O. Olatayo
Department of Mathematical Science,
Olabisi Onabanjo University, Ago-Iwoye,
Nigeria

7

ABSTRACT

This paper introduced a bootstrap method called truncated geometric bootstrap method for time series stationary process. We estimate the parameters of a geometric distribution which has been truncated as a probability model for the bootstrap algorithm. This probability model was used in resampling blocks of random length, where the length of each blocks has a truncated geometric distribution. The method was able to determine the block sizes b and probability p attached to its random selections. The mean and variance were estimated for the truncated geometric distribution and the bootstrap algorithm developed based on the proposed probability model.

INTRODUCTION

Several studies have been made of truncated distribution. [1] has treated the truncated binomial distribution. [2] [3] also treated the truncated binomial and negative binomial distribution and truncated Poisson distribution and has shown how to estimate the parameter of distributions. The first part of the present paper shows how to estimate the parameter of truncated geometric distribution as a true probability model, by method of moment. The second part of the paper uses the distribution to developed bootstrap algorithm for stationary time series process to overcome the difficulties of moving block scheme [4] [5] and geometric Bootstrap scheme of [6] in determining probability p and block size b respectively. The heart of the bootstrap is not simply computer simulation, and bootstrapping is not perfectly synonymous with Monte Carlo. Bootstrap method relies on using an original sample or some part of it, such as residuals as an artificial population from which to randomly resample [7]. Bootstrap resampling methods have emerged as powerful tools for constructing inferential procedures in modern statistical data analysis. The Bootstrap approach, as initiated by [8] , avoids having to derive formulas via different analytical arguments by taking advantage of fast computers. The bootstrap methods have and will continue to have a profound influence throughout science; as the availability of fast, inexpensive computing has enhanced our abilities to make valid statistical inferences about the world without the need for using unrealistic or unverifiable assumptions [9].

An excellent introduction to the bootstrap may be found in the work of [10] -[13]. Recently, [4] and [5] have independently introduced nonparametric versions of the bootstrap that are applicable to weakly dependent stationary observations. Their sampling procedure have been generalized by[6] by resampling "blocks of blocks" of observations for the stationary time series process. On this note we propose a stationary bootstrap method generated by resampling blocks of random size, where the length of each block has a "truncated geometric distribution".

MATERIALS AND METHOD

The truncation is more than just theoretical interest as a number of applications have been reported in [14].

Estimating the Parameter of Truncated Geometric Distribution.

A random variable X may be defined to have a truncated geometric distribution, with parameter p. and N terms, when it has the probability distribution.

$$P(X=x)=k(1-P)^{x-1}P \quad (x=1,2,\cdots,N) \tag{2.1}$$

The constant K is found, using condition

$$\sum P(X=x)=1,$$

to be

$$1/\left[1-(1-P)^{N}\right].$$

We shall introduce a truncated form of the geometric distribution and then demonstrate that it is suitable model for a probability distribution. Hence, the probability density function is:

$$\sum_{j=0}^{\alpha} P(1-P)^{jN+r-1} = P(1-P)^{r-1}\sum_{j=0}^{\alpha}(1-P)^{jN} \tag{2.2}$$

$$=\frac{P(1-P)^{r-1}}{1-(1-P)^{N}} \tag{2.3}$$

(r=1, 2,…, N)

These are the probabilities of truncated geometric distribution with parameter P and N terms. Therefore we intend to derive the distributional properties of our estimates by looking at the expectation and variance of truncated geometric distribution.

Thus,

$$P(r)=\frac{P(1-P)^{r-1}}{1-(1-P)^{N}}$$

$$E(r)=\sum_{r=1}^{N} rP(r)=\sum_{r=1}^{N}\frac{P}{1-(1-P)^{N}}r(1-P)^{r-1}=\frac{P}{1-q^{N}}\sum_{r=1}^{N} rq^{r-1}\quad (q=(1-p)) \tag{2.4}$$

Recall some facts on geometric series,

$$q^{1}+q^{2}+\cdots+q^{N}=\frac{q\left(1-q^{N}\right)}{1-q} \tag{2.5}$$

Using method of moment we have;

$$\frac{d}{dq}\left[\frac{q(1-q)^{N}}{1-q}\right]=\frac{d}{dq}\left[\frac{q-q^{N+1}}{1-q}\right]=\frac{1-q^{N}+q-Nq^{N}(1-q)}{(1-q)^{2}} \tag{2.6a}$$

$$\frac{p}{1-q^{N}}\sum_{r=1}^{N} rq^{r-1}=\frac{1-q}{1-q^{N}}\left[\frac{1-q^{N}+q-Nq^{N}(1-q)}{(1-q)^{2}}\right] \tag{2.6b}$$

$$E(r)=\frac{1-q^{N}+q-Nq^{N}(1-q)}{\left(1-q^{N}\right)(1-q)} \tag{2.7}$$

Thus, E(r) is the sample mean, as demonstrated by [15].

In this paper, we derived the second moment further.

$$V(r)=\sum_{r=1}^{N} r^{2}p(r)-\left(\sum_{r=1}^{N} rp(r)\right)^{2} \tag{2.8}$$

Thus,

$$\sum_{r=1}^{N} r^{2}p(r)=\sum_{r=1}^{N} r\frac{p(1-p)^{r-1}}{1-(1-P)^{N}}$$

$$\frac{1-q}{1-q^N}\left[\sum_{r=1}^{N} r(r-1)(-p)^{r-1}+\sum_{r=1}^{N} r(1-p)^{r-1}\right]$$

For the second moment and by second differentiation we have,

$$\frac{1-q}{1-q^N}\left[q\left(\frac{d}{dq}\left(\frac{1-q^N-Nq^N(1-q)}{(1-q)^2}\right)+\frac{1-q^N-Nq^N(1-q)}{(1-q)^2}\right)\right]$$

$$E\left(r^2\right)=\frac{(1+q)-q^N\left((1-q)^2 N(N+1)+(1+q)(1+N(1-q))\right)}{(1-q)^2\left(1-q^N\right)} \qquad (2.9)$$

For the variance

$$V\left(r^2\right)=\frac{(1+q)-q^N\left((1-q)^2 N(N+1)+(1+q)(1+N(1-q))\right)}{(1-q)^2\left(1-q^N\right)}-\left[\frac{1-q^N-Nq^N(1+q)}{(1-q)\left(1-q^N\right)}\right]^2$$

The variance for truncated geometric distribution is:

$$\Rightarrow\frac{\left(1+q^{2N}\right)q-q^N\left(1+q^2\right)N^2+q^{N+1}\left(N^2-1\right)}{(1-q)^2\left(1-q^N\right)^2}$$

Therefore, when N = 0 we have $\frac{q}{p}$ and $\frac{q}{p^2}$ corresponding to the mean and variance of geometric distribution.

Suppose that length L, numbered 1 to N are to be selected randomly with replacement which is the principle of bootstrap as introduced by [8]. Then the process continues until it is truncated geometrically at r with an appropriate probability p attached to its random selections in form of r, r+1, r+2, L, r+N−1.

RESULTS AND DISCUSSIONS

In this paper, we take our N to be 4, that is, the random selections could be truncated between 1 to 4 at an appropriate probability attached to each of them respectively.

Using Equation (2.3) and from any fixed value of p in (0, 1). Different values of p can be generated for any values of r. Therefore our r here will represent the number of blocks to be selected and the appropriate values of p attached to it are the probabilities of being selected at each random selection with replacement.

Below are the table value generated for a number of fixed number of p in (0, 1), for different values of rth truncation geometrically.

From the Table 1, each value r = 1 to N will be our b (block sizes) and each values of p (probability) for each various random selections.

Table 1: Summary Statistics values for truncated geometric distribution

P	R = 1	R = 2	R = 3	R = 4
0.1	0.2908	0.2617	0.2356	0.2120
0.2	0.3388	0.2710	0.2168	0.1734
0.3	0.3949	0.2764	0.1935	0.1354
0.4	0.4596	0.2757	0.1654	0.0993
0.5	0.5333	0.2667	0.1333	0.0667
0.6	0.6158	0.2463	0.0985	0.0394
0.7	0.7057	0.2117	0.0635	0.0191
0.8	0.8013	0.1603	0.0321	0.0064
0.9	0.9001	0.0900	0.0090	0.0009

Suppose that L length, numbers 1 to N are to be selected randomly with replacement. Then, the process continues until it is truncated geometrically at r with an appropriate probability P attached to its random selection in form of r, r+1, r+2, L, r+N−1. We take our N to be 4, that is, the random selections could be truncated between 1 to 4 at an appropriate probability.

Then, a description of the resampling algorithm when r > 1 is as follows:

1) Let X_1, L, X_N be a random variables 2) Let X_1^* be determined by the r-th truncated observation X_r in the original time series 3) Let X_{i+1}^* be equal to X_{r+1} with probability 1 − P and picked at random from the original N observations with probability p. 4) Let $B_{i,b}=\{X_i, X_{i+1}, L, X_{i+b-1}\}$, be the block consisting of b observation starting from X_i 5) Let B_{I_1,L_1}, B_{I_2,L_2}, L be a sequence of blocks of random length determined by truncated geometric distribution. 6) The first L_1, observations in the pseudo time series X_1^*, X_2^*, L, X_N^* are determined by the first block B_{I_1,L_1} of observation $X_{I_1}, L, X_{I_1+L_{i-1}}$; the next L_2 observations in the pseudo time series are the observations in the second sampled block $X_{I_2+L_{2-1}}$.. 7) The process is resampled with replacement, until the process is stopped once N observation in the pseudo time series have been generated. 8) Once X_1^*, L, X_N^* has been generated, one can compute the quantities of interest for the pseudo time series.

The algorithm has two major components, the construction of a bootstrap sample and the computation of statistics on this bootstrap sample, and repeats these operation many times through some kinds of a loop.

Proposition 1. Conditional on $X_1, X_2, L, X_N, X_1^*, X_2^*, L, X_N^*$ is stationary.

If the original observations X_1, L, X_N are all distinct, then the new series X_1^*, L, X_N^* is, conditional on X_1, L, X_N a stationary Markov chain. If, on the other hand, two of the original observations are identical and the remaining are distinct, then the new series X_1^*, L, X_N^* is a stationary second order Markov chain. The stationary bootstrap resampling scheme proposed here is distinct from the proposed by [4] [5] , but posses the same properties with that proposed by [2].

CONCLUSIONS

The truncated geometric bootstrap method proposed in this paper is a true probability model with the estimated parameters. The method was able to determine the block sizes b and probability p attached to its random selections. The mean and variance were estimated for the truncated geometric distribution and the bootstrap algorithm developed based on the proposed probability model.

REFERENCES

1. Finney, D.J. (1949) The Truncated Binomial Distribution. Annals of Eugenics, 14, 319-328. http://dx.doi.org/10.1111/j.1469-1809.1947.tb02410.x
2. Rider, P.R. (1953) Truncated Poisson Distribution. Journal of the American Statistical Association, 48, 826-830. http://dx.doi.org/10.1080/01621459.1953.10501204
3. Rider, P.R. (1955) Truncated Binomial and Negative Binomial Distribution. Journal of the American Statistical Association, 50, 877-883.http://dx.doi.org/10.1080/01621459.1955.10501973
4. Kunsch, H.R. (1989) The Jacknife and the Bootstrap for General Stationary Observations. The Annals of Statistics, 17, 1217-1241. http://dx.doi.org/10.1214/aos/1176347265
5. Liu, R.Y. and Singh, K. (1992) Moving Blocks Jackknife and Bootstrap Capture Weak Dependence. In: R. Lepage and L. Billard, Eds., Exploring the Limits of Bootstrap, John Wiley, New York.
6. Politis, D.N. and Romano, J.O. (1994) The Stationary Bootstrap. Journal of American Statistical Association, 89, 303-1313.http://dx.doi.org/10.1080/01621459.1994.10476870
7. Barreto, H. and Howland, F.M. (2005) Introductory Econometrics, Using Monte Carlo simulation with Microsoft Excel. Cambridge University Press, Cambridge. http://dx.doi.org/10.1017/CBO9780511809231
8. Efron, B. (1979) Bootstrap Methods: Another Look at the Jacknife. The Annals of Statistics, 7, 1-26. http://dx.doi.org/10.1214/aos/1176344552
9. Leger, C., Politis, D. and Romano, J. (1992) Bootstrap Technology and Applications. Technometrics, 34, 378-398. http://dx.doi.org/10.1080/00401706.1992.10484950

10. Efron, B. and Tibshirani, R. (1986) Bootstrap Measures for Standard Errors, Confidence Intervals, and Other Measures of Statistical Accuracy. Statistical Science, 1, 54-77.http://dx.doi.org/10.1214/ss/1177013815
11. Efron, B. and Tibshirani, R. (1993) An Introduction to the Bootstrap. Chapman and Hall/CRC, London
12. Diciccio, T. and Romano, J. (1988) A Review of Bootstrap Confidence Intervals (with Discussion). Journal of the Royal Statistical Society B, 50, 338-370.
13. Davison, A.C. and Hinkley, D.V. (1997) Bootstrap Methods and their Application. Cambridge University Press, Cambridge. http://dx.doi.org/10.1017/CBO9780511802843
14. Bancroft, G.A., Colwell, D.J. and Gillet, J.R. (1983) A Truncated Poisson Distribution. The Mathematical Gazette, 66, 216-218.
15. Kapadia, C.H. and Thomasson, R.L. (1975) On Estimating the Parameter of the Truncated Geometric Distribution by the Method of Moments. Annals of the Institute of Statistical Mathematics, 20, 519-532.

CITATION

Olatayo, T. (2014) Truncated Geometric Bootstrap Method for Time Series Stationary Process. Applied Mathematics, 5, 2057-2061. doi: 10.4236/am.2014.513199.

Reliability Estimators for the Components of Series and Parallel Systems: the Weibull Model

Felipe L. Bhering[1], *Carlos A. de B. Pereira*[1], *and Adriano Polpo*[2]
[1]Department of Statistics, University of São Paulo, São Paulo, Brazil
[2]Department of Statistics, Federal University of São Carlos, São Carlos, Brazil

8

ABSTRACT

This paper presents a hierarchical Bayesian approach to the estimation of components' reliability (survival) using a Weibull model for each of them. The proposed method can be used to estimation with general survival censored data, because the estimation of a component's reliability in a series (parallel) system is equivalent to the estimation of its survival function with right- (left-) censored data. Besides the Weibull parametric model for reliability data, independent gamma distributions are considered at the first hierarchical level for the Weibull parameters and independent uniform distributions over the real line as priors for the parameters of the gammas. In order to evaluate the model, an example and a simulation study are discussed.

INTRODUCTION

This paper presents a hierarchical Bayesian approach to the estimation of components' reliability using a Weibull model for each component in series and parallel systems. A series system is a frame of components that works if and only if all its components are functional, that is, whenever one fails the system fails. As a dual frame, the parallel system fails if and only if all components are malfunctioning.

The literature dealing with the problem of estimating the reliability of series systems, or competing risks, is abundant. The work of Kaplan and Meier [1] is arguably the most celebrated work, where it was developed a nonparametric estimator using a frequentist approach. For a Bayesian counterpart, we draw the attention to Salinas-Torres et al. [2] and Polpo and Sinha [3]. Rodrigues et al. [4] performed a simulation study of three different methods to estimate the reliability of a series system. They compared the Kaplan-Meier estimator [1] , maximum likelihood estimator (MLE) and the Bayesian plug-in estimator (BPE) for Weibull reliability systems. Their results indicated that MLE and BPE are similar in quality and that both outperformed the Kaplan-Meier estimator. However, the construction of credible bounds was not addressed in their work.

For parallel systems, the literature is scarce. To the best of our knowledge, Polpo and Pereira [5] were the first to address the nonparametric reliability estimation in parallel systems and their components, using the Bayesian paradigm. Later on, Bhattacharya and Samaniego [6] proposed a frequentist nonparametric estimator for components' reliability under a restrictive condition that all components are independent and identically distributed.

In a related work, Polpo et al. [7] presented the reliability estimation with Weibull models and non-informative priors, also using the Bayesian paradigm. Their proposal had very demanding computational needs, and the described Markov Chain Monte Carlo (MCMC) algorithm suffered from convergence issues in many problem instances, making the estimation a difficult task. The authors realized later on that it is often impossible to evaluate the components' reliability with that method because of such issues, and unfortunately one cannot predict when the algorithm will succeed. The present work aims at achieving a robust estimation procedure for the estimation problem. We suggest considering non-informative priors in a one-level hierarchical model as means to solve the estimation issues faced by the algorithm of Polpo et al. [7]. Moreover, an important goal of this work is to provide a simple way to build credible bounds for the reliability functions, giving a step-by-step algorithm that performs such analysis.

Similarly to Polpo et al. [7] and Rodrigues et al. [4], we consider Weibull statistical models for the system's reliability. Two independent gamma distributions are considered in the first hierarchical level. The gamma distributions are parametrized by their means and variances, instead of the more common parametrization with scale and shape. In the second level of the hierarchy, we choose two flat priors for the means, and fixed values for the variances of the gamma distributions in the first level. These hyper-parameters corresponding to variances in the first level can be seen as prior precision parameters. The means of the gamma distributions (first level of the hierarchy) can be viewed as the prior expectations of the parameters of the Weibull model. In this work, the posterior modes are taken as the Bayesian estimators of the gamma distributions.

The estimation of the reliability functions has three main steps: 1) we draw a sample from the posterior distribution of the Weibull parameters; 2) using the appropriate transformation, we build a sample from the reliability posterior distribution; and 3) locally, for each reliability time, we evaluate the posterior mean. The high posterior density (HPD) procedure was used to define the credible region for the reliability function. We emphasize that we are not using the plug-in estimator, but the posterior mean of the reliability function, which seems more suitable under the Bayesian paradigm.

This paper is organized as follows. Section 2 describes all functions that are involved in the estimation procedure. Section 3 provides the estimation procedure itself. Section 4 presents a simulation study that highlights the quality of the model and the proposed estimators, and final remarks and additional comments are given in Section 5. We note that an extended abstract of this work has appeared in the Brazilian Conference on Bayesian Statistics [8].

THE MODEL

We use the same notation as in Polpo et al. [7]. Consider a system of k components and let X_j, j, be the sequence of failure times of all components. We assume that this sequence is composed of independent random variables with a (possibly) different Weibull distribution for each component. Recall that we only observe a random vector of two variables, namely, (T,δ) with $T=\min(X_1,\ldots,X_k)$ for the series system and $T=\max(X_1,\ldots,X_k)$ for the parallel system, with $\delta=j$ if $T=X_j$, for $j=1,\ldots,k$. The δ quantity can be viewed as an indicator function of the component that caused the system failure.

Consider a sample of n independent and identically distributed systems (either all series or all parallel systems). The observations are represented by

$$(T,\delta)=\{(T_i,\delta_i):i=1,\cdots,n\}$$

The reliability function of the jth component is given by $R_j(t)=P(X_j>t)$, $j=1,\ldots,k$. Therefore, the reliability function is $R(t)=\prod_{j=1}^{k}R_j(t)$ for the series system and $R(t)=1-\prod_{j=1}^{k}(1-R_j(t))$ for the parallel system.

We define random variables X_j for the components' reliability with Weibull distributions parametrized by $\theta_j=(\beta_j,\eta_j)$, that is,

$$P(X_j>x|\theta_j)=R(x|\theta_j)=\exp\left\{-\left(\frac{x}{\eta_j}\right)^{\beta_j}\right\} \tag{1}$$

For $x>0$, $\beta_j>0$ (shape) and $\eta_j>0$ (scale). Then, the likelihood function for the series system is given by

$$L(\theta|t,\delta)\propto\prod_{j=1}^{k}\prod_{i=1}^{n}\left[f_j(t_i|\theta_j)\right]^{\mathbb{I}_{\{\delta_i=j\}}}\left[R_j(t_i|\theta_j)\right]^{1-\mathbb{I}_{\{\delta_i=j\}}}, \tag{2}$$

and for the parallel system,

$$L(\theta|t,\delta) \propto \prod_{j=1}^{k}\prod_{i=1}^{n}\left[f_j(t_i|\theta_j)\right]^{\mathbb{I}\{\delta_i=j\}}\left[1-R_j(t_i|\theta_j)\right]^{1-\mathbb{I}\{\delta_i=j\}}, \tag{3}$$

where f is the density function of a random variable with Weibull distribution, $\theta = (\theta_1,\ldots,\theta_k)$, and $_A$ is the indicator function of the set A.

The prior distributions were considered independent with

$$\beta_j \sim \text{gamma}\left(m_{\beta_j}, v_{\beta_j}\right), \quad \eta_j \sim \text{gamma}\left(m_{\eta_j}, v_{\eta_j}\right)$$

$$\pi\left(m_{\beta_j}\right) \propto \pi\left(m_{\eta_j}\right) \propto 1$$

and v_{β_j} and v_{η_j} as known constants, j=1,…,k. Then

$$\pi(\boldsymbol{\vartheta}) \propto \pi\left(\boldsymbol{\theta} \mid \boldsymbol{m}_\beta, \boldsymbol{v}_\beta, \boldsymbol{m}_\eta, \boldsymbol{v}_\eta\right)\pi\left(\boldsymbol{m}_\beta\right)\pi\left(\boldsymbol{v}_\beta\right)\pi\left(\boldsymbol{m}_\eta\right)\pi\left(\boldsymbol{v}_\eta\right)$$

$$\propto \prod_{j=1}^{k}\pi\left(\theta_j \middle| m_{\beta_j}, v_{\beta_j}, m_{\eta_j}, v_{\eta_j}\right)\pi\left(m_{\beta_j}\right)\pi\left(v_{\beta_j}\right)\pi\left(m_{\eta_j}\right)\pi\left(v_{\eta_j}\right)$$

$$\propto \prod_{j=1}^{k}\frac{\beta_j^{m_{\beta_j}^2/v_{\beta_j}-1}\exp\left\{-m_{\beta_j}\beta_j/v_{\beta_j}\right\}}{\left(v_{\beta_j}/m_{\beta_j}\right)^{m_{\beta_j}^2/v_{\beta_j}}\Gamma\left(m_{\beta_j}^2/v_{\beta_j}\right)}\frac{\eta_j^{m_{\eta_j}^2/v_{\eta_j}-1}\exp\left\{-m_{\eta_j}\eta_j/v_{\eta_j}\right\}}{\left(v_{\eta_j}/m_{\eta_j}\right)^{m_{\eta_j}^2/v_{\eta_j}}\Gamma\left(m_{\eta_j}^2/v_{\eta_j}\right)},$$

where

$$\boldsymbol{\vartheta} = \left(\boldsymbol{\theta}, \boldsymbol{m}_\beta, \boldsymbol{v}_\beta, \boldsymbol{m}_\eta, \boldsymbol{v}_\eta\right)$$

.

$$\boldsymbol{m}_{\beta} = \left(m_{\beta_1}, \cdots, m_{\beta_k}\right)$$

and

$$\boldsymbol{m}_{\eta} = \left(m_{\eta_1}, \cdots, m_{\eta_k}\right)$$

are the prior mean parameters,

$\mathrm{v}_{\beta} = \left(\mathrm{v}_{\beta_1}, \ldots, \mathrm{v}_{\beta_k}\right)$ and $\mathrm{v}_{\eta} = \left(\mathrm{v}_{\eta_1}, \ldots, \mathrm{v}_{\eta_k}\right)$ are the prior variance (precision) parameters, and

$$m_{\beta_j}, m_{\eta_j}, v_{\beta_j}, v_{\eta_j} > 0,$$

j=1,..., k l.

In this case, we have that the posterior distributions of series and parallel systems are, respectively,

$$\pi(\boldsymbol{\vartheta} | \boldsymbol{t}, \boldsymbol{\delta}) \propto \pi(\boldsymbol{\vartheta}) \prod_{j=1}^{k} \prod_{i=1}^{n} \left[\frac{t_i^{\beta_j - 1} \beta_j}{\eta_j} \exp\left\{ -\left(\frac{t_i}{\eta_j} \right)^{\beta_j} \right\} \right]^{\mathbb{I}\{\delta_i = j\}} \left[\exp\left\{ -\left(\frac{t_i}{\eta_j} \right)^{\beta_j} \right\} \right]^{1 - \mathbb{I}\{\delta_i = j\}},$$

and

$$\pi(\boldsymbol{\vartheta} | \boldsymbol{t}, \boldsymbol{\delta}) \propto \pi(\boldsymbol{\vartheta}) \prod_{j=1}^{k} \prod_{i=1}^{n} \left[\frac{t_i^{\beta_j - 1} \beta_j}{\eta_j} \exp\left\{ -\left(\frac{t_i}{\eta_j} \right)^{\beta_j} \right\} \right]^{\mathbb{I}\{\delta_i = j\}} \left[1 - \exp\left\{ -\left(\frac{t_i}{\eta_j} \right)^{\beta_j} \right\} \right]^{1 - \mathbb{I}\{\delta_i = j\}},$$

where $t=(t_1,\ldots,t_n)$ are the observed failure time of the system, $\delta = (\delta_1, \ldots, \delta_n)$ are the indicators of which component failed, and the other quantities are as defined before.

ESTIMATION

For the estimation, we use the EM algorithm [9] to obtain the posterior mode as estimates of m_β and m_η, and the MCMC procedure to generate a sample from the posterior distribution of β and η. The algorithm steps for the estimation are briefly given as follows:

1. Choose the prior precision values v_β and v_η. Note that one can set to the same value all the precision values, that is, $v = v_\beta = v_\eta$. We suggest the use of $v=4$.
2. Choose the initial guess for the parameters to be estimated: m_β, m_η, β and η.
3. Using the initial guess, consider m_β and m_η as fixed values. Employing the MCMC tool, generate a sample (of size n_p) from the posterior distribution of β and η. We suggest the use of $n_p=1000$. It may also be necessary to use a "burn-in" and a "jump" to ensure convergence of the MCMC.
4. (Expectation step of the EM) Using the posterior sample of β and η obtained in Step 3, evaluate the mean of the likelihood function, obtaining a "mean" function of m_β and m_η.
5. (Maximization step of the EM) Find the values of m_β and m_η that maximize the function obtained in Step 4.
6. Update the initial guess of m_β and m_η with the values obtained in Step 5, and the values of β and η with their posterior mean obtained in Step 3.
7. Repeat Steps 3, 4, 5 and 6 until convergence of m_β and m_η is reached. We suggest the use of a tolerance value of three decimal places between the previous values of m_β and m_η and the current ones, in order to decide whether to stop iterating.
8. Once convergence of m_β and m_η is reached, use their values to generate a sample from the posterior distribution of β and η by applying the MCMC tool.

Using this algorithm, we obtain the posterior mode (the Bayesian estimate) of m_β and m_η, and a sample (of size n_p) from the joint posterior distribution of β and. If we estimate the reliability function of any component (let us arbitrarily choose component 1), then we can notice that the estimations of the other components' reliability functions are very similar and could be omitted here. Consider that the sample from the posterior of the parameters of component 1 can be expressed as $(\beta_{11}, \beta_{12}, \ldots, \beta_{1n_p})$ and $(\eta_{11}, \eta_{12}, \ldots, \eta_{1n_p})$. To obtain the reliability estimates and credible regions, consider the functions $Y_\ell(t) = F(t|\beta_{1\ell}, \eta_{1\ell})$, $\ell = 1, \ldots, n_p$, where $F(t|\beta_{1\ell}, \eta_{1\ell})$ are Weibull distribution functions conditioned on $\beta_{1\ell}$ and $\eta_{1\ell}$. Consequently, the posterior mean estimate of the component's reliability can be expressed as

$$\widehat{R(t)} = \mathrm{E}\left[R(t)\middle|\text{data}\right] = 1 - \frac{1}{n_p}\sum_{\ell=1}^{n_p} Y_\ell(t), \quad \text{for each } t.$$

Hence, for each fixed t, $Y(t) = (Y_1(t), \ldots, Y_{n_p}(t))$ is a sample from the posterior of the component 1's distribution function and, to obtain the credible region, we can either use the quantiles of Y(t) or evaluate the high posterior density credible interval. To estimate the mean reliability time, we have

$$\mathrm{E}\left[T\middle|\text{data}\right] = \frac{1}{n_p}\sum_{\ell=1}^{n_p} \mathrm{E}\left[T\middle|\beta_{1\ell}, \eta_{1\ell}\right], \quad \ell = 1, \cdots, n_p,$$

where

$$\mathrm{E}\left[T\middle|\beta_{1\ell}, \eta_{1\ell}\right] = \eta_{1\ell}\Gamma\left(1 + 1/\beta_{1\ell}\right)$$

is the mean of a random variable with Weibull distribution, and is the gamma function. Note that similar procedures can be used to evaluate other quantities of interest; as an example, the posterior median could be evaluated.

EXAMPLES

Example 1

Consider three random variables X_1, X_2, X_3 such that X_1 has Weibull distribution with mean 2 and variance 4, X_2 has gamma distribution with mean 2 and variance 0.667, and X_3 has log-normal distribution with mean 2.014 and variance 6.968. We have generated a sample (with size n = 100) of series systems with these three components and another sample (again with size n = 100) of parallel systems with the exactly same three components. The components were chosen in order to have similar means but different variances and, consequently, different distributions. We have used the same theoretical components in both simulations (series and parallel systems) to verify, in each situation, the differences that are due to the distinct system models with the available data. The simulated data have the following characteristics: 1) for the series systems, we have obtained 64%, 80%, and 56% of censured data for components 1, 2, and 3, respectively; and 2) for the parallel systems, we have observed 61%, 68% and 71%, respectively for the same three components. In this case, the main interest is in the estimation of the components' reliability functions. Note that, with our simulated example, we have a huge amount of censored data, making it a challenging example.

As already said, the estimation procedures are performed using MCMC. We have discarded the first 10,000 samples (as burn-in) from the posterior to achieve the stationary measure and then have generated a sample from the posterior. To perform the estimation of the reliability functions and the credible region, we have used a sample of size 1000 from the posterior, which was obtained by discarding 10 samples (the jump between each final sample point). We have used v=4 in the prior specification for all parameters and systems. For the experiment with series systems, we have obtained $\widehat{m_\beta}$ =(1.31,4.12,1.44) and $\widehat{m_\eta}$ =(2.19,2.03,1.80). For the parallel systems, the estimates are $\widehat{m_\beta}$ =(1.28,2.57,0.92) and $\widehat{m_\eta}$ =(2.67,2.28,2.11). To evaluate the quality of these estimates, we have compared the "true" reliability of each component with the estimated reliability function. Table 1 presents the posterior mean and the posterior standard deviation of each parameter involved in the model (for both series and parallel systems). Note that the standard deviations are relatively small, and the estimation of the mean reliability time is very close to the original values, indicating a good performance of our method.

It can be seen from the 95% credible bounds that the "true" reliability of each component was well estimated. We note however that the "true" reliability functions of the components are, for short (time) intervals, outside the 95% credible bounds. Considering that these are reasonably challenging examples, this situation is likely to happen in any estimation procedure (see Figure 1and Figure 2).

Example 2

This example has a simulation study to show the quality of the proposed hierarchical Weibull model over many different conditions. We have considered 108 different scenarios that were built using three different sample sizes (n=30,100,1000), three different proportions of censored data (0%, 20% and 40%), two different means (2 and 7) of reliability time for the generating distribution, three different generating distributions (Weibull, gamma, log-normal), and two types of censoring (right and left). In all these cases, we have fixed the variance of the generating distribution to 5. We have used a non-random censored approach to guarantee the desired proportions of censure, that is, for each scenario, we have fixed a time for which all values that are larger than this time are assumed censored for the right-censored data (series systems), and all values smaller than this fixed time are assumed censored for the left-censored data (parallel systems). In order to improve the analysis, 100 copies of each scenario were considered.

To summarize and to compare the results of the simulation, we have evaluated the bias and the mean squared error (MSE) of the estimated mean reliability time, for each scenario. The results are presented in

Tables 2-7. One can see from the results that all biases and MSEs are close to zero, indicating that, in all scenarios, the model has estimated well the true mean reliability time. Even in the most challenging scenarios, which are those with small sample size (n=30) and large proportion of censoring (40%), the estimated mean reliability time has had a good performance. When the generating distribution is Weibull, the model should obviously estimate well, yet the results show that the performances were good even for the other two generating (non-Weibull) distributions.

Table 1: Posterior mean (sd) of some quantities involved in the estimation of the simulated examples

	Series system estimates		
	β	η	$E(T\|\beta,\eta)$
Component 1	1.26 (0.17)	2.27 (0.41)	2.13 (0.45)
Component 2	3.98 (0.53)	2.06 (0.12)	1.87 (0.11)
Component 3	1.40 (0.17)	1.83 (0.22)	1.68 (0.22)
	Parallel system estimates		
	β	η	$E(T\|\beta,\eta)$
Component 1	1.23 (0.15)	2.60 (0.27)	2.45 (0.22)
Component 2	2.47 (0.32)	2.25 (0.14)	2.00 (0.13)
Component 3	0.87 (0.11)	1.98 (0.33)	2.17 (0.29)

FINAL REMARKS

We have introduced a Bayesian reliability statistical analysis using hierarchical models for the problem of estimating the reliability functions and credible bounds of series and parallel systems. The MCMC has shown good performance in terms of convergence, making the inference process simple and efficient. It shall be noted that this performance is not dependent on our choice of a "non-informative" scheme to define the prior hyper-parameters. This is important because other researchers may want to fairly compare our method with other frequentist estimators. However, informative priors may very well produce additional improvements in the estimates. The Example 1 has shown good robustness in the sense that the model has performed well for all components in both series and parallel systems. Another important aspect is that we can obtain credible bounds for the reliability function, task that is usually hard if one uses a plug-in estimator for the reliability function. The Example 2 provides an extensive simulation study with more than one hundred different scenarios. Overall, the model has performed very well for estimating the mean reliability time. Some open questions that should be addressed in future works are the development of hypothesis tests for the components, for instance, one can have interest in testing the hypothesis of equal means of all components (or a subset of components), and the extension of these ideas to more general coherent systems.

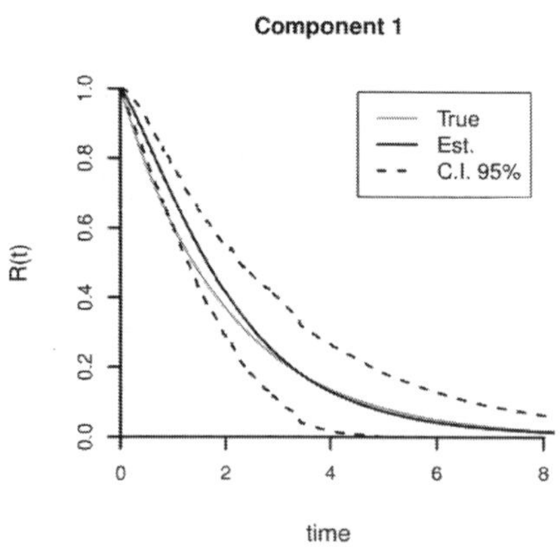

(a)

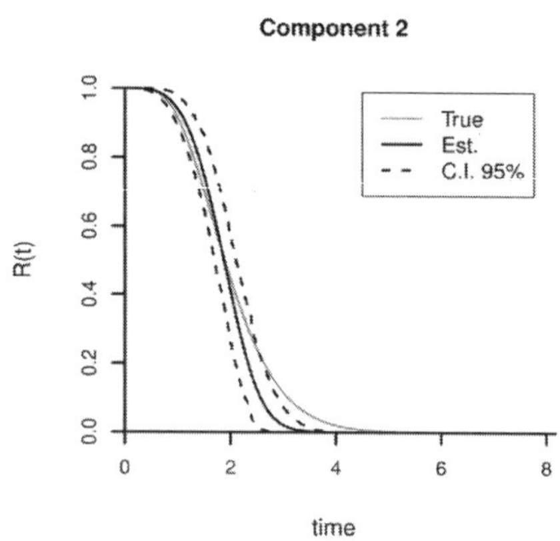

(b)

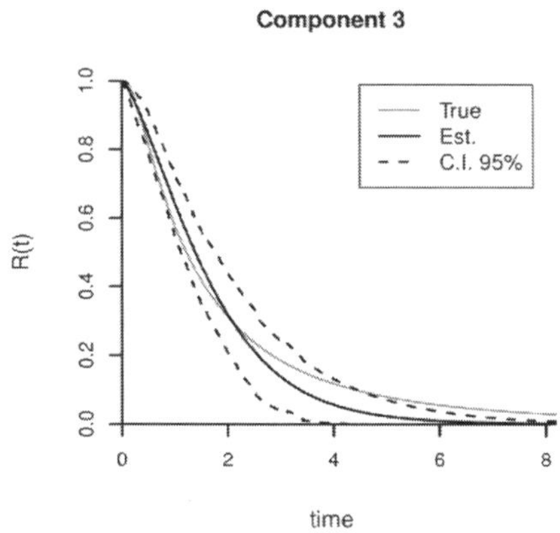

(c)

Figure 1: Reliability of the components in the experiment with series systems: (a) Component 1; (b) Component 2; (c) Component 3.

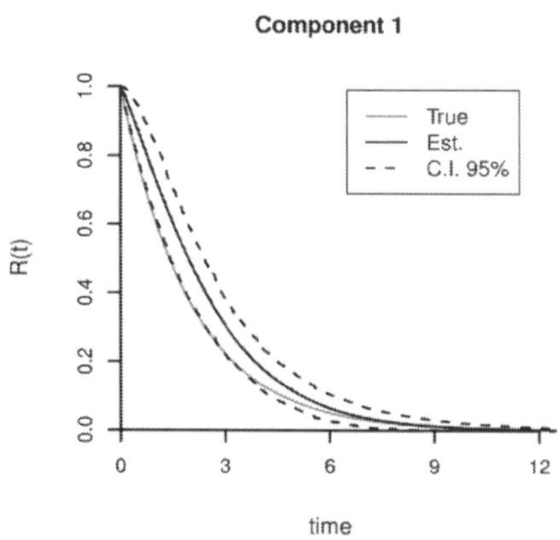

(a)

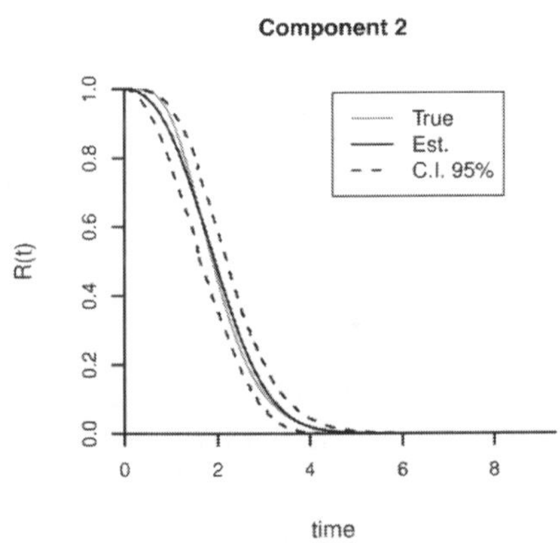

(b)

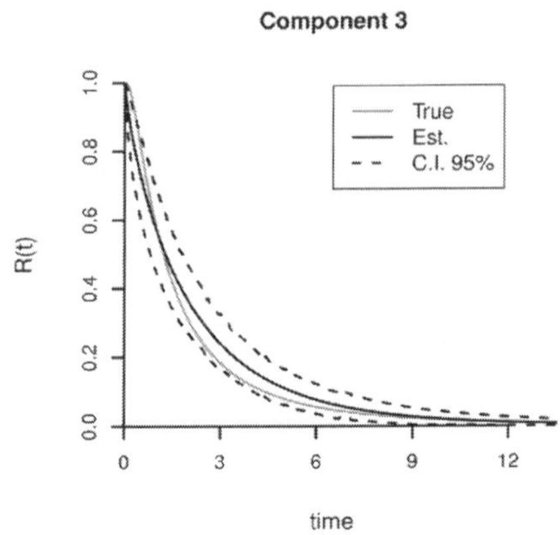

(c)

Figure 2: Reliability of the components in the experiment with parallel systems: (a) Component 1; (b) Component 2; (c) Component 3.

Table 2: Bias and mean squared error (MSE) of the E(T) estimate for data generated from the Weibull distribution with right-censored data

Cens.	True E(T)	n = 30		n = 100		n = 1000	
		Bias	**MSE**	**Bias**	**MSE**	**Bias**	**MSE**
0%	2	−0.1300	0.2217	−0.0426	0.0572	−0.0057	0.0043
0%	7	0.0293	0.1523	0.0026	0.0474	−0.0095	0.0043
20%	2	−0.2813	0.4893	−0.1242	0.1275	−0.0200	0.0068
20%	7	−0.0522	0.2307	−0.0233	0.0565	−0.0148	0.0047
40%	2	−0.4293	0.7364	−0.2394	0.2925	−0.0239	0.0108
40%	7	−0.1773	0.3590	−0.0661	0.0874	−0.0162	0.0055

Table 3: Bias and mean squared error (MSE) of the E(T) estimate for data generated from the gamma distribution with right-censored data

Cens.	True E(T)	n = 30 Bias	MSE	n = 100 Bias	MSE	n = 1000 Bias	MSE
0%	2	−0.1407	0.2286	−0.0191	0.0593	−0.0003	0.0055
0%	7	0.0006	0.1845	0.0081	0.0623	0.0098	0.0048
20%	2	−0.3829	0.6346	−0.1132	0.1234	−0.0560	0.0111
20%	7	0.0569	0.1721	0.1121	0.0669	0.1049	0.0153
40%	2	−0.5068	0.9424	−0.3084	0.3807	−0.1174	0.0299
40%	7	0.1389	0.2505	0.2372	0.1107	0.2336	0.0582

Table 4: Bias and mean squared error (MSE) of the E(T) estimate for data generated from the log-normal distribution with right-censored data

Cens.	True E(T)	n = 30 Bias	MSE	n = 100 Bias	MSE	n = 1000 Bias	MSE
0%	2	−0.1160	0.1757	−0.0024	0.0396	−0.0089	0.0050
0%	7	−0.0348	0.1551	0.0462	0.0429	0.0220	0.0052
20%	2	0.0720	0.1410	0.2475	0.0779	0.2523	0.0655
20%	7	0.0786	0.1492	0.1848	0.0637	0.1620	0.0296
40%	2	0.2373	0.1795	0.4205	0.1981	0.4371	0.1928
40%	7	0.2190	0.2094	0.3438	0.1532	0.3337	0.1147

Table 5: Bias and mean squared error (MSE) of the E(T) estimate for data generated from the Weibull distribution with leftcensored data

Cens.	True E(T)	n = 30 Bias	MSE	n = 100 Bias	MSE	n = 1000 Bias	MSE
0%	2	−0.1300	0.2217	−0.0426	0.0572	−0.0057	0.0043
0%	7	0.0293	0.1523	0.0026	0.0474	−0.0095	0.0043
20%	2	0.1030	0.1178	0.0034	0.0475	−0.0060	0.0043
20%	7	0.0387	0.1562	−0.0001	0.0490	−0.0094	0.0043
40%	2	−0.1813	0.2660	−0.0471	0.0579	−0.0065	0.0043
40%	7	0.0301	0.1813	0.0040	0.0597	−0.0099	0.0048

Table 6: Bias and mean squared error (MSE) of the E(T) estimate for data generated from the gamma distribution with leftcensored data

Cens.	True E(T)	n = 30		n = 100		n = 1000	
		Bias	MSE	Bias	MSE	Bias	MSE
0%	2	−0.1407	0.2286	−0.0192	0.0594	−0.0003	0.0055
0%	7	0.0006	0.1845	0.0081	0.0623	0.0098	0.0048
20%	2	0.1029	0.1038	0.0448	0.0490	0.0008	0.0054
20%	7	0.1108	0.2143	0.0950	0.0739	0.0932	0.0139
40%	2	−0.1934	0.2463	−0.0245	0.0602	−0.0029	0.0055
40%	7	0.2185	0.2842	0.2105	0.1242	0.1850	0.0402

Table 7: Bias and mean squared error (MSE) of the E(T) estimate for data generated from the log-normal distribution with left-censored data

Cens.	True E(T)	n = 30		n = 100		n = 1000	
		Bias	MSE	Bias	MSE	Bias	MSE
0%	2	−0.1160	0.1757	−0.0024	0.0396	−0.0089	0.0050
0%	7	−0.0348	0.1551	0.0462	0.0429	0.0220	0.0052
20%	2	−0.0092	0.1253	0.0254	0.0405	0.0211	0.0052
20%	7	0.1003	0.1701	0.1757	0.0784	0.1375	0.0240
40%	2	−0.1165	0.1937	0.0575	0.0441	0.0624	0.0086
40%	7	0.2359	0.2727	0.3138	0.1548	0.2763	0.0825

REFERENCES

1. Kaplan, E.L. and Meier, P. (1958) Nonparametric Estimation from Incomplete Observations. Journal of the American Statistical Association, 53, 457-481.http://dx.doi.org/10.1080/01621459.1958.10501452
2. Salinas-Torres, V., Pereira, C.A.B. and Tiwari, R. (2002) Bayesian Nonparametric Estimation in a Series System or a Competing Risks Model. Journal of Nonparametric Statistics, 14, 449-458. http://dx.doi.org/10.1080/10485250213114
3. Polpo, A. and Sinha, D. (2011) Correction in Bayesian Nonparametric Estimation in a Series System or a Competing-Risks Model. Statistics & Probability Letters, 810, 1756-1759. http://dx.doi.org/10.1016/j.spl.2011.07.023

4. Rodrigues, A.S., Dias, T.C.M., Lauretto, M. and Polpo, A. (2012) Reliability Analysis in Series Systems: An Empirical Comparison between Bayesian and Classical Estimators. 31st International Workshop on Bayesian Inference and Maximum Entropy Methods in Science and Engineering, Waterloo, AIP Conference Proceedings, 1443, 214-221.
5. Polpo, A. and Pereira, C.A.B. (2009) Reliability Nonparametric Bayesian Estimation in Parallel Systems. IEEE Transactions on Reliability, 580, 364-373.http://dx.doi.org/10.1109/TR.2009.2019493
6. Bhattacharya, D. and Samaniego, F.J. (2010) Estimating Component Characteristics from System Failure-Time Data. Naval Research Logistics (NRL), 570, 380-389.
7. Polpo, A., Coque, M.A. and Pereira, C.A.B. (2009) Statistical Analysis for Weibull Distributions in Presence of Right and Left Censoring. The Proceedings of 8th International Conference on Reliability, Maintainability and Safety (ICRMS), 1, 219-223.
8. Bhering, F., Polpo, A. and Pereira, C.A.B. (2012) A Hierarchical Weibull Bayesian Model for Series and Parallel Systems. AIP Conference Proceedings, 1490, 59-66.http://dx.doi.org/10.1063/1.4759589
9. McLachlan, G.J. and Krishnan, T. (2008) The EM Algorithm and Extensions. John Wiley & Sons, Hoboken. http://dx.doi.org/10.1002/9780470191613

CITATION

Bhering, F. , Pereira, C. and Polpo, A. (2014) Reliability Estimators for the Components of Series and Parallel Systems: The Weibull Model. Applied Mathematics, 5, 1633-1640. doi: 10.4236/am.2014.511157.

Estimation for Nonnegative First-Order Autoregressive Processes with an Unknown Location Parameter

Andrew Bartlett and William McCormick
University of Georgia, Athens, USA

9

ABSTRACT

Consider a first-order autoregressive processes $X_t = \phi X_{t-1} + Z_t$, where the innovations are nonnegative random variables with regular variation at both the right endpoint infinity and the unknown left endpoint $\boldsymbol{\theta}$. We propose estimates for the autocorrelation parameter f and the unknown location parameter $\boldsymbol{\theta}$ by taking the ratio of two sample values chosen with respect to an extreme value criteria for f and by taking the minimum of $X_t - \hat{\phi}_n X_{t-1}$ over the observed series, where $\hat{\phi}_n$ represents our estimate for f. The joint limit distribution of the proposed estimators is derived using point process techniques. A simulation study is provided to examine the small sample size behavior of these estimates.

INTRODUCTION

In many applications, the desire to model the phenomena under study by non-negative dependent processes has increased. An excellent presentation of the classical theory concerning these models can be found, for example, in Brockwell and Davis [1]. Recently, advancements in such models have shifted focus to some specialized features of the model, e.g. heavy tail innovations or nonnegativity of the model. In this paper we examine the behavior of traditional estimates under conditions leading to non-Gaussian limits. For example, the standard approach to parameter estimation within the AR(1) process is through the Yule-Walker estimator;

$$\hat{\phi}_{LS} = \frac{\sum_{t=1}^{n-1}(X_t - \bar{X})(X_{t+1} - \bar{X})}{\sum_{t=1}^{n}(X_{t+1} - \bar{X})^2}, \tag{1.1}$$

Where $\overline{X} = \frac{1}{n}\sum_{t=1}^{n} X_{t^*}$.

A slightly different approach presented in Mathew and McCormick [2] used linear programming to obtain estimates for ϕ and θ under certain optimization constraints. While there are many established methods to estimate the autocorrelation coefficient in an AR(1) model, there are just a few approaches on estimating the unknown location parameter in an AR(1) model. One was mentioned in McCormick and Mathew [2] where they considered

$$\hat{\theta}_{\text{range}} = X_{j^*} - \hat{\phi}_{\text{range}} X_{j^*-1}, \text{ where } \hat{\phi}_{\text{range}} = \frac{X_{t^*+1} - X_{j^*}}{X_{t^*} - X_{j^*-1}},$$

t*and j* provides the index of the maximal and minimal X_i respectively for $1 \le i \le n$.

In this paper we examine estimation questions and asymptotic properties of alternative estimates for ϕ and θ respectively, relating to the model

$$X_t = \phi X_{t-1} + Z_t, t \ge 1,$$

where $0 < \phi < 1, \theta > 0$ and $\{Z_t\}$ is an i.i.d. sequence of nonnegative random variables whose innovation distribution F is assumed to be regularly varying at infinity with index $-\beta$ and regularly varying at θ with index α, where θ, denotes the unknown but positive left endpoint. As a result of not restricting the innovations $\{Z_t\}$ to be bounded on a finite range, we can first estimate the autoregressive parameter ϕ through regular variation at infinity and then estimate the positive but unknown location parameter through regular variation at θ, the left endpoint.

While we have mentioned a few established estimation procedures, one notable exception was that of maximum likelihood. Although typically intractable and intricate in the time series setting, when the innovations in the AR(1) model are exponential, the maximum likelihood procedure had a major contribution on the estimation of positive heavy tailed time series. With these considerations in mind, Raftery [3] determined the limiting distribution of the maximum likelihood estimate for the autocorrelation coefficient ϕ. As a result, the estimator

$$\hat{\phi}_n = \bigwedge_{t=1}^{n} \frac{X_t}{X_{t-1}}, \tag{1.2}$$

was considered. The realization of this estimator was the stepping stone for the work done in this paper along with Davis and McCormick [4] which first considered this alternative estimator and used a point process approach to obtain the asymptotic distribution of the natural estimator $\hat{\phi}_n$. This was done in the context that the innovations distribution F varies regularly at 0, the left endpoint, and satisfy some moment condition.

The work presented in this paper is an extension of the work done in Davis and McCormick [4] including the following contributions to dependent time series with heavy-tail innovations. The first contribution involves the development of estimates for the autocorrelation coefficient and unknown location parameter under regular variation at both endpoints, with a rate of convergence $n^{1/\beta}\ell(n)$, where ℓ is slowly varying function. The second contribution involves using an extreme value method, e.g. point processes to establish the asymptotic distribution of the proposed estimators and weak convergence for the asymptotically independent joint distribution. An initial observation is that our estimation procedure is especially easy to implement for both ϕ and θ. That is, the autoregressive coefficient ϕ in the causal AR(1) process is estimated by taking the minimum of the ratio of two sample values while estimation for the unknown location parameter θ was achieved through minimizing $X_t - \hat{\phi}_n X_{t-1}$ over the observed series.

This naturally motivates a comparison between the estimation procedure presented in this paper and the standard linear programming estimates mentioned above, since within a nonnegative AR(1) model the linear programming estimate reduces to the estimate proposed, namely, $\min_{1\le t\le n}(X_t / X_{t-1})$, where $\{X_t\}$ denotes the AR(1) process. This comparison along with the comparison between Mathew and McCormick's [2] optimization method and Bartlett and McCormick [5] extreme value method was performed through simulation and is presented in Section 3. The results found appear to demonstrate a favorable performance for our extreme value method over the 3 alternative estimators.

The main proofs in this paper rely heavily on point process methods from extreme value theory. The essential idea is to first establish the convergence of a sequence of point processes based on simple quantities and then apply the continuous mapping theorem to obtain convergence of the desired statistics. More background information on point processes, regular variation, and weak convergence can be found in Resnick [6]. Also, a nice survey on linear programming estimation procedures and nonnegative time series can be found in Anděl [7], Anděl [8], and Datta and McCormick [9], whereas more applications on modeling the phenomena with heavy tailed distributions and ensuing estimation issues can be found in Resnick [10].

The rest of the paper is organized as follows: asymptotic limit results for the autocorrelation parameter ϕ, unknown location parameter θ, and joint distribution of (ϕ,θ) are presented in Section 2, while Section 3 is concerned with the small sample size behavior of these estimates through simulation.

ASYMPTOTICS

The following point process limit result is fundamental. Since the result makes no use of an ARMA structure, we present it for more general linear models subject to usual summability conditions on the coefficients. In that regard for this result, we assume that $\{X_n, n \geq 0\}$ is the stationary linear process given by

$$X_n = \sum_{j=0}^{\infty} c_j Z_{n-j}$$

with $\sum_{j=0}^{\infty} |c_j|^{\delta} < \infty$ for some $0 < \delta < \beta, \delta \leq 1$. Furthermore for this result we may relax our assumptions on the innovation distribution and we require that $|Z_1|$ has a regularly varying tail distribution, i.e.,

$P(|Z_1| > x) = x^{-\beta}\ell(x), x > 0$ for a slowly varying function ℓ and the innovation distribution is tail balanced

$$\frac{P(Z_1 > x)}{P(|Z_1| > x)} \to p \text{ and } \frac{P(Z_1 \leq -x)}{P(|Z_1| > x)} \to q \text{ as } x \to \infty.$$

Define point processes

$$\xi_n = \sum_{k=1}^{n} \epsilon_{\left(b_n^{-1} X_{k-1}, Z_k\right)}, n \geq 1$$

and let $\sum_{k\geq 1} \epsilon_{j_k}$ denote PRM (v) on $\mathbb{R}_0 = \mathbb{R} / \{0\}$ where

v has Radon-Nikodym derivative with respect to Lebesgue measure

$$\frac{d\nu}{d\lambda}(x) = p\beta x^{-\beta-1} 1_{(0,\infty)}(x) + q\beta |x|^{-\beta-1} 1_{(-\infty,0)}(x).$$

Let $\{Z_{i,k}, i0, k \geq 1\}$ be an iid array with $Z_{i,k} \overset{d}{=} Z_1$ and independent of $\sum_{k=1}^{\infty} \varepsilon_{j_k}$. Define

$$\xi = \sum_{k=1}^{\infty} \sum_{i=0}^{\infty} \epsilon_{\left(c_i j_k, Z_{i,k}\right)}$$

Our basic result is to show that $M_p(\mathbb{R}_0 \times \mathbb{R})$ equipped with the topology of vague convergence

$$\xi_n \xrightarrow{d} \xi \text{ in}$$

which is close in statement and spirit to Theorem 2.4 in Davis and Resnick [11]. In view of the commonality of the two results, we present only the needed changes to the Davis and Resnick proof to accommodate the current setting. Aside from keeping track of the time when points occur, i.e. large jumps, the difference in the point processes considered here with those in Davis and Resnick [11] is the inclusion of marks, i.e. the second component of the point $(b_n^{-1} X_{k-1}, Z_k)$. This complication induces an additional weak dependence in the points which is addressed in Lemma 2.2 through a straight forward blocking argument. First, we establish weak convergence of marked point processes of a normalized vector of innovations. For a positive integer m define

$$Z_k^{(m)} = \left(Z_k, Z_{k-1}, \cdots, Z_{k-m+1}\right)$$

and point process

$$\mathcal{I}_n^{(m)} = \sum_{k=1}^{n} \epsilon_{\left(b_n^{-1} Z_k^{(m)}, Z_{k+1}\right)}, n \geq 1.$$

Let $e_i = (0,\cdots,1,\cdots,0), 1 \leq i \leq m$ denote the standard basis vectors for $\mathbb{R}^m$. Define an associated marked point process with the first component placed on an axis by

$$\tilde{\mathcal{I}}_n^{(m)} = \sum_{k=1}^{n} \sum_{i=1}^{m} \epsilon_{\left(b_n^{-1} Z_k e_i, Z_{k+i}\right)}$$

In the following Lemma, we show that $I_n^{(m)}$ and $\tilde{I}_n^{(m)}$ are asymptotically indistinguishable in the following sense. Let $\mathbb{R}_0^m = \mathbb{R}^m / \{0\}$ and $E = \mathbb{R}_0^m \times \mathbb{R}$. Consider the class of rectangles

$$\mathcal{S} = \left\{ R \times (-\infty, x] : R = \mathop{\times}_{i=1}^{m} (c_i, d_i], \right.$$
$$\left. x < \infty \text{ and } \bar{R} \times (-\infty, x] \subset E \right\}$$

Lemma 2.1: As n tends to infinity

$$\mathcal{I}_n^{(m)}(B) - \tilde{\mathcal{I}}_n^{(m)}(B) \xrightarrow{P} 0 \text{ for all } B \in \mathcal{S}.$$

Proof: Following the proof presented in Proposition 2.1 of Davis and Resnick [11], suppose that $B \in S$ is such that for some $1 \leq i' \leq m, B \cap (\mathbb{R}e_{i'} \times \mathbb{R}) \neq \phi$. As noted in Davis and Resnick [11] for all $i \neq i'$, one has $c_i < 0 < d_i$. Observe that

$$\mathcal{I}_n^{(m)}(B) \le \mathcal{I}_n^{(m)}\left(\mathbb{R}\times\cdots\times[c_{i'},d_{i'}]\times\cdots\times\mathbb{R}\times(-\infty,x]\right)$$
$$= \sum_{k=-i'+2}^{n-i'+1} \epsilon_{\left(b_n^{-1}Z_k e_{i'},Z_{k+i'}\right)}\left(\mathbb{R}\times\cdots\times[c_{i'},d_{i'}]\times\cdots\times\mathbb{R}\times(-\infty,x]\right) \tag{2.1}$$

Similarly let $c = \vee_{i\neq i'} c_i < 0$ and $d = \wedge_{i\neq i'} d_i > 0$. Then

$$\mathcal{I}_n^{(m)}(B)$$
$$\ge \mathcal{I}_n^{(m)}\left([c,d]\times\cdots\times[c_{i'},d_{i'}]\times\cdots\times[c,d]\times(-\infty,x]\right)$$
$$\ge \mathcal{I}_n^{(m)}\left(\mathbb{R}\times\cdots\times[c_{i'},d_{i'}]\times\cdots\times\mathbb{R}\times(-\infty,x]\right)$$
$$-\sum_{i\neq i'}\mathcal{I}_n^{(m)}\left(E_i\times(-\infty,x]\right) \tag{2.2}$$

where $y=(y_1,\cdots,y_m)\in E_i$ according to $y_i \notin [c,d]$ and $y_{i'}\in[c_{i'},d_{i'}]$. Note that

$$E\left[\mathcal{I}_n^{(m)}\left(E_i\times(-\infty,x]\right)\right]$$
$$\le nP\left(\frac{1}{b_n}|Z_1|>|c|\wedge d,\frac{1}{b_n}|Z_1|>|c_{i'}|\wedge|d_{i'}|\right)$$
$$=o(1). \tag{2.3}$$

Thus from (2.1)-(2.3) we obtain

$$E\left|\mathcal{I}_n^{(m)}(B)-\tilde{\mathcal{I}}_n^{(m)}(B)\right|$$
$$\le 2(i'-1)P\left(\frac{1}{b_n}Z_1>|c_{i'}|\wedge|d_{i'}|\right)+o(1)=o(1)$$
$$\text{as } n\to\infty. \tag{2.4}$$

Then the result follows as in Davis and Resnick [11], Proposition 2.1, completing the proof.

Lemma 2.2: Let $V_n^{(m)}$ and $V^{(m)}$ be the point processes on the space $E = \mathbb{R}_0 \times \mathbb{R}^m$ defined by

$$V_n^{(m)} = \sum_{k=1}^{n} \in_{(b_n^{-1}Z_k, Z_{k+m}^{(m)})} \text{ and } V^{(m)} = \sum_{k=1}^{\infty} \in_{(j_k, Y_k^{(m)})}$$

where $\{Y_k^{(m)}, k \geq 1\}$ is an iid sequence with

$Y_1^{(m)} \overset{d}{=} (Z_1, \cdots, Z_m)$ and is independent of $\sum_{k=1}^{\infty} \in_{j_k}$. Then in $M_P(E)$,

$$\mathcal{V}_n^{(m)} \xrightarrow{d} \mathcal{V}^{(m)}.$$

Proof: We employ a blocking argument to establish this result. Let r_n be a sequence of integers such that $r = r_n \to \infty$ as $n \to \infty$ and $r = 0(n)$. Let $h = \lfloor n/r \rfloor$ and $l \geq m$. Define blocks

$$I_{n,s} = [r(s-1)+1, \cdots, rs-l],$$
$$J_{n,s} = [rs-l+1, \cdots, rs]$$

for $1 \leq s \leq h$ and $J_{n,h+1} = [rh+1, \cdots, n]$.

Then it is clear that for $s \neq t$

$$\sigma(Z_{k+i}, k \in I_{n,s}, 0 \leq i \leq m)$$
$$\perp\!\!\!\perp \sigma(Z_{k+i}, k \in I_{n,t}, 0 \leq i \leq m).$$

Write

$$\mathcal{V}_{n,s}^{(m)} = \sum_{k \in I_{n,s}} \epsilon_{\left(b_n^{-1} Z_k , Z_{k+m}^{(m)}\right)} \quad \text{and} \quad \tilde{\mathcal{V}}_{n,t}^{(m)} = \sum_{k \in J_{n,s}} \epsilon_{\left(b_n^{-1} Z_k , Z_{k+m}^{(m)}\right)}$$

Then

$$\mathcal{V}_n^{(m)} = \sum_{s=1}^{h} \mathcal{V}_{n,s}^{(m)} + \sum_{t=1}^{h+1} \tilde{\mathcal{V}}_{n,t}^{(m)}. \tag{2.5}$$

Let $B \subset E$ be a disjoint union of rectangles

$$B = \bigcup_{i=1}^{j} B_i \tag{2.6}$$

where $B_i = [c_i, d_i] \times R_i$ with $R_i = X_{l=1}^{m}[x_{il}, y_{il}]$. Let $\mu = v \times F \times \cdots \times F$ denote the mean measure of $V^{(m)}$ which is $PRM(\mu)$ on$E$. To complete the proof we first show that for all sets B of the form given in (2.6) that

$$\lim_{n \to \infty} P\left(\mathcal{V}_n^{(m)}(B) = 0\right) = \exp\left(-\mu(B)\right). \tag{2.7}$$

The above limit result follows from the easily verifiable relations:

$$P\left(\mathcal{V}_{n,1}^{(m)}(B) = \cdots = \mathcal{V}_{n,h}^{(m)}(B) = 0\right)$$
$$= P^h\left(\mathcal{V}_{n,1}^{(m)}(B) = 0\right); \tag{2.8}$$

$$P\left(\mathcal{V}_{n,s}^{(m)}(B_k) \wedge \mathcal{V}_{n,s}^{(m)}(B_l) \geq 1\right) = O\left(\frac{r^2}{n^2}\right)$$

for $1 \leq k \neq l \leq j$; (2.9)

$$P\left(\mathcal{V}_{n,s}^{(m)}\left(\bigcup_{i=1}^{j} B_i\right) \geq 1\right)$$

$$= \sum_{i=1}^{j} P\left(\mathcal{V}_{n,s}^{(m)}(B_i) \geq 1\right) + O\left(\frac{r^2}{n^2}\right); \tag{2.10}$$

$$P\left(\mathcal{V}_{n,1}^{(m)}(B_i) \geq 1\right) = \frac{r}{n}\mu(B_i)(1+o(1)); \tag{2.11}$$

and

$$\sum_{t=1}^{h+1} \tilde{\mathcal{V}}_{n,t}^{(m)}(B) \xrightarrow{p} 0 \text{ as } n \to \infty. \tag{2.12}$$

Indeed, in view of (2.5) and (2.12), (2.7) is equivalent to showing

$$\lim_{n\to\infty} P\left(\sum_{s=1}^{h} \mathcal{V}_{n,s}^{(m)}(B) = 0\right) = \exp(-\mu(B)) \tag{2.13}$$

and the above relation holds by (2.8), (2.10), and (2.11), viz.

$$P\left(\sum_{s=1}^{h} \mathcal{V}_{n,s}^{(m)}(B) = 0\right) = \left(1 - P\left(\mathcal{V}_{n,1}^{(m)}(B) \geq 1\right)\right)^h$$

$$= \left(1 - \sum_{i=1}^{j} \frac{r}{n}\mu(B_i) + o(r/n)\right)^h$$

$$\to \exp\left(-\sum_{i=1}^{j} \mu(B_i)\right).$$

It is immediate that for a rectangle

$B = [c,d] \times X_{i=1}^{m}[x_i, y_i] \subset E$ we have

$$\lim_{n\to\infty} E\left[\mathcal{V}_n^{(m)}(B)\right] = \mu(B). \tag{2.14}$$

Therefore the result is seen to hold by (2.7) and (2.14) by application of Theorem 4.7 in Kallenberg [12].

Lemma 2.3: Let $I_n^{(m)}$ and $I^{(m)}$ be point processes on the space $E = \mathbb{R}_0^{(m)} \times \mathbb{R}$

$$I_n^{(m)} = \sum_{k=1}^{n} \in_{(b_n^{-1}Z_k^{(m)}, Z_{k+1})}, n \geq 1 \text{ and } I^{(m)} = \sum_{k=1}^{\infty}\sum_{i=1}^{m} \in_{(j_k e_i, z_{i,k})}$$

Then in $M_p(E)$,

$$I_n^{(m)} \xrightarrow{d} I^{(m)}$$

Proof: We begin by applying the argument used in Theorem 2.2 of Davis and Resnick [11] with the modification that the relevant composition of maps of point processes is given by

$$\sum_{k\geq 1} \epsilon_{(u_k, v_{mk}, \cdots, v_{1k})} \mapsto \left(\sum_{k\geq 1} \epsilon_{(u_k, v_{mk})}, \ldots, \sum_{k\geq 1} \epsilon_{(u_k, v_{1k})}\right)$$

$$\mapsto \left(\sum_{k\geq 1} \epsilon_{(u_k e_1, v_{1k})}, \ldots, \sum_{k\geq 1} \epsilon_{(u_k e_m, v_{mk})}\right)$$

$$\mapsto \sum_{k>1}\sum_{i=1}^{m} \epsilon_{(u_k e_i, v_{ik} \cdot)}$$

Each map being continuous, the composition is a continuous map from $M_p(\mathbb{R}_0 \times \mathbb{R}^m)$ to $M_p(\mathbb{R}_0^{(m)} \times \mathbb{R})$ with each space being equipped with the topology of vague convergence. Therefore by the continuous mapping theorem and Lemma 2.2 we obtain

$$\tilde{\mathcal{I}}_n^{(m)} = \sum_{k=1}^{\infty}\sum_{i=1}^{m} \epsilon_{\left(b_n^{-1} Z_k e_i, Z_{k+i}\right)} \xrightarrow{d} \sum_{k=1}^{\infty}\sum_{i=1}^{m} \epsilon_{\left(j_k \mathbf{e}_i, Z_{i,k}\right)}$$

$$= \mathcal{I}^{(m)}. \tag{2.15}$$

Finally we complete the proof by Lemma 2.1 and (2.15) arguing as in Davis and Resnick [11].

We are now ready to present our fundamental result.

Theorem 2.1: Let ξ_n and ξ be the point processes on the space $E = \mathbb{R}_0 \times \mathbb{R}$ defined by

$$\xi_n = \sum_{k=1}^{n} \epsilon_{\left(b_n^{-1} X_{k-1}, Z_k\right)}, n \geq 1 \text{ and } \xi = \sum_{k=1}^{\infty}\sum_{i=0}^{\infty} \epsilon_{\left(c_i j_k, Z_{i,k}\right)}$$

where $\sum_{k=1}^{\infty} \epsilon_{j_k}$ is PRM(v) and $\{Z_{i,k}, I \geq 0, k \geq 1\}$ is an iid array with $Z_{i,k} \stackrel{d}{=} Z_1$ and independent of $\sum_{k=1}^{\infty} \epsilon_{j_k}$.

Then in $M_p(E)$,

$$\xi_n \xrightarrow{d} \xi.$$

Remark. Apart from considering a time coordinate and restricting the process to an AR (1) process, the above Theorem 2.1 and Theorem 3.1 in Mathew and McCormick [2] consider essentially the same point process limit result. However, their result gave a wrong limit point process. This error is corrected in the current paper.

Proof: Observe that the map

$$\left(z_k, z_{k-1}, \cdots, z_{k-m+1}\right) \mapsto \sum_{i=0}^{m-1} c_i z_{k-i}$$

induces a continuous map on point processes given by

$$\sum_{k=1}^{\infty} \epsilon_{\left(z_k, z_{k-1}, \cdots, z_{k-m+1}, z_{k+1}\right)} \mapsto \sum_{k=1}^{\infty} \epsilon_{\left(\sum_{j=0}^{m-1} c_j z_{k-j}, z_{k+1}\right)}.$$

Thus we obtain from Lemma 2.3 that

$$\sum_{k=1}^{n} \epsilon_{\left(b_n^{-1} \sum_{i=0}^{m-1} c_i Z_{k-i}, Z_{k+1}\right)} \xrightarrow{d} \sum_{k=1}^{\infty} \sum_{i=0}^{m-1} \epsilon_{\left(c_i j_k, Z_{i,k}\right)}. \tag{2.16}$$

The result now follows from (2.16) by the same argument in Davis and Resnick [11] to finish their Theorem 2.4.

Returning to the AR(1) model under discussion in this paper and the estimate $\hat{\phi}_n$ given in (1.2), we obtain the following asymptotic limit result.

Theorem 2.2. Let $\{X_t, t \geq 0\}$ be the stationary solution to the AR (1) recursion $X_t = \phi X_{t-1} + Z_t$ and

$\hat{\phi}_n = \wedge_{t=1}^{n} \dfrac{X_t}{X_{t-1}}$. Under the assumptions that $0 < \phi < 1$ and the innovation distribution F has regularly varying right tail with index $-\beta$ and finite positive left endpoint θ,

$$\lim_{n\to\infty} P\left(b_n\left(\hat{\phi}_n-\phi\right)>x\right)=\mathrm{e}^{-x^{\beta}EW^{-\beta}} \text{ for all } x>0,$$

where $W=\wedge_{i=0}^{\infty}\frac{Z_i}{\phi^i}$ and $b_n=F^{\leftarrow}(1-1/n)$.

Proof: For $x>0$ define a subset

$$Q_x=\left\{(x_1,x_2):\frac{x_2}{x_1}\le x, x_1>0, x_2\ge\theta\right\}.$$

Then note that for the point processes

$\xi_n=\sum_{k=1}^{n}\epsilon_{(b_n^{-1}X_{k-1},Z_k)}$, we have

$$\{\xi_n(Q_x)=0\}=\left\{\bigwedge_{k=1}^{n}\frac{Z_k}{b_n^{-1}X_{k-1}}>x\right\}=\left\{b_n\left(\hat{\phi}_n-\phi\right)>x\right\}.$$

Applying Theorem 2.1 in the case of an AR(1) process so that $c_i=\phi^i, i\ge 0$, we have

$$\xi_n \xrightarrow{d} \xi=\sum_{k=1}^{\infty}\sum_{i=0}^{\infty}\epsilon_{\left(\phi^i j_k, Z_{i,k}\right)}.$$

Note that as a subset of $E=(0,\infty]\times[\theta,\infty)$, the set Q_x is a bounded continuity set with respect to the limit point process ξ so that

$$\lim_{n\to\infty} P\left(b_n\left(\hat{\phi}_n-\phi\right)>x\right)=P\left(\xi\left(Q_x\right)=0\right)$$

$$=P\left(\bigwedge_{k=1}^{\infty}\bigwedge_{i=0}^{\infty}\frac{Z_{i,k}}{\phi^i j_k}>x\right)=P\left(\bigwedge_{k=1}^{\infty}\frac{W_k}{j_k}>x\right)$$

where $(W_k=\wedge_{i=0}^{\infty}\phi, Z_{i,k}, k\ge 1)$ is an i.i.d. sequence independent of $(j_k, k\ge 1)$. Let

$$\tilde{\xi}=\sum_{k\ge 1}\epsilon_{(j_k, W_k)}. \tag{2.17}$$

Then by Proposition 5.6 in Resnick [10], we have that if G denotes the distribution of W_1, then $\tilde{\xi}$ is a Poisson random measure on E with mean measure $\mu = v\times G$, where $c(dx)=\beta x^{-\beta-1}{}_{[0,\infty)}(x)dx$. Using (2.17) we can write

$$P\left(\bigwedge_{k=1}^{\infty}\frac{W_k}{j_k}>x\right)=P\left(\tilde{\xi}\left(Q_x\right)=0\right)=\exp\left(-\mu\left(Q_x\right)\right).$$

Since $\mu(Q_x)=x^{\beta}EW^{-\beta}$, the result follows.

Corollary 2.1: Under conditions given in Theorem 2.2,

$$\hat{\phi}_n \xrightarrow{a.s} \phi.$$

Proof: Since $b_n = F^{\leftarrow}(1-1/n)\to\infty$ we have

$\hat{\phi}_n \xrightarrow{p} \phi$. But this implies $\hat{\phi}_n \xrightarrow{a,s} \phi$. since $\hat{\phi}_n \ge \phi$ and is non-increasing.

Let us now define our estimate of θ:

$$\hat{\theta}_n = \bigwedge_{t \in I_n} \left(X_t - \hat{\phi}_n X_{t-1} \right),$$

where we define the index set

$I_n = \{t : 1 \le t \le n \text{ and } X_{t-1} \le (a_n b_n)^p\}$ where $0 < p < 1$

is a fixed value.

Lemma 2.4: Under the assumptions that F is regularly varying with index α at its positive left endpoint θ and $\overline{F}$ is regularly varying with index $-\beta$ at infinity, its right endpoint, and $\alpha > \beta$, then

$$a_n^{-1} \left(\hat{\theta}_n - \bigwedge_{t \in I_n} Z_t \right) \xrightarrow{P} 0, \text{as } n \to \infty,$$

Where $a_n = F^{\leftarrow}(1/n) - \theta$. Furthermore, for any $y \ge 0$

$$\lim_{n\to\infty} P\left(a_n^{-1}\left(\hat{\theta}_n - \theta\right) > y\right)$$

$$= \lim_{n\to\infty} P\left(a_n^{-1} \left(\bigwedge_{t=1}^{n} Z_t - \theta \right) > y \right) = e^{-y^\alpha}.$$

Proof: Since $\alpha > \beta$, we have $\lim_{n\to\infty} a_n b_n = \infty$.

Therefore since $(b_n(\hat{\phi} - \phi), n \ge 1)$ is a tight sequence by Theorem 2.2 and since $\max_{t \in I_n} X_{t-1} / (a_n b_n)^p \le 1$ with $0 < p < 1$, we have

$$a_n^{-1} \left(\hat{\phi}_n - \phi \right) \bigvee_{t \in I_n} X_{t-1} \xrightarrow{P} 0.$$

The first statement now follows since

$$a_n^{-1}\left|\hat{\theta}_n - \bigwedge_{t\in I_n} Z_t\right| \le a_n^{-1}\left|\hat{\phi}_n - \phi\right| \bigvee_{t\in I_n} X_{t-1}.$$

For the second statement observe

$$0 \le P\left(a_n^{-1}\left(\bigwedge_{t\in I_n} Z_t - \theta\right) > y\right) - P\left(a_n^{-1}\left(\bigwedge_{t=1}^{n} Z_t - \theta\right) > y\right)$$

$$= P\left(a_n^{-1}\left(\bigwedge_{t=1}^{n} Z_t - \theta\right) \le y < a_n^{-1}\left(\bigwedge_{t\in I_n} Z_t - \theta\right)\right)$$

$$\le P\left(\bigcup_{1\le t\le n}\left(X_{t-1} > (a_n b_n)^{\rho} \text{ and } a_n^{-1}(Z_t - \theta) \le y\right)\right)$$

$$\le nP\left(X_0 > (a_n b_n)^{\rho}\right)P(Z_1 \le a_n y + \theta) = o(1). \tag{2.18}$$

The result for the second statement now follows from (2.18) and the first part of the lemma. Finally, the identification of the limit distribution is well known.

A useful observation follows from this lemma which we state as a corollary.

Corollary 2.2: Under the assumptions of Lemma 2.4 for any $x, y > 0$

$$\lim_{n\to\infty}\left\{Pb_n\left(\left(\hat{\phi}_n - \phi\right) > x, a_n^{-1}\left(\hat{\theta}_n - \theta\right) > y\right)\right.$$

$$\left. - P\left(b_n\left(\hat{\phi}_n - \phi\right) > x, a_n^{-1}\left(\bigwedge_{t=1}^{n}(Z_t - \theta) > y\right)\right)\right\} = 0$$

Proof: By Lemma 2.4 we have

$$\lim_{n\to\infty}\left\{P\left(b_n\left(\hat{\phi}_n-\phi\right)>x, a_n^{-1}\left(\hat{\theta}_n-\theta\right)>y\right)\right.$$

$$\left.-P\left(b_n\left(\hat{\phi}_n-\phi\right)>x, a_n^{-1}\left(\bigwedge_{t\in I_n} Z_t-\theta\right)>y\right)\right\}=0$$

The result then follows from

$$0\le E1\left\{b_n\left(\hat{\phi}_n-\phi\right)>x\right\}$$

$$\cdot\left(1\left\{a_n^{-1}\left(\bigwedge_{t\in I_n} Z_t-\theta\right)>y\right\}-1\left\{a_n^{-1}\left(\bigwedge_{t=1}^{n} Z_t-\theta\right)>y\right\}\right)$$

$$\le P\left(a_n^{-1}\left(\bigwedge_{t\in I_n} Z_t-\theta\right)>y\right)-P\left(a_n^{-1}\left(\bigwedge_{t=1}^{n} Z_t-\theta\right)>y\right).$$

Corollary 2.2 allows a simplification in determining the joint asymptotic behavior of $(\hat{\phi}_n, \hat{\theta}_n)$ by allowing us to replace $\hat{\theta}_n$ with $\min_{1\le t\le n} Z_t$. The next lemma will provide another useful simplification—this time on $\hat{\phi}_n$.

For a positive integer m define

$$X_t^{(m)}=\sum_{i=0}^{m-1}\phi^i Z_{t-i}.$$

Lemma 2.5: Let $U_n^{(m)}$ and U_n be defined as

$$U_n^{(m)} = \bigwedge_{t=1}^{n} \frac{Z_t}{b_n^{-1} X_{t-1}^{(m)}} \quad \text{and} \quad U_n = \bigwedge_{t=1}^{n} \frac{Z_t}{b_n^{-1} X_{t-1}}.$$

Then for any $\in > 0$

$$\lim_{m\to\infty} \lim_{n\to\infty} P\left(\left|U_n^{(m)} - U_n\right| > \epsilon\right) = 0.$$

Proof: We first note for any positive M that

$$P\left(\left|U_n^{(m)} - U_n\right| > \epsilon\right) = P\left(U_n^{(m)} U_n \left|\frac{1}{U_n^{(m)}} - \frac{1}{U_n}\right| > \epsilon\right)$$

$$\leq P\left(\left|\frac{1}{U_n^{(m)}} - \frac{1}{U_n}\right| > \epsilon / M^2\right) + P\left(U_n^{(m)} > M\right).$$

In order to calculate $P\left(\left| \frac{1}{U_n^{(m)}} - \frac{1}{U_n} \right| > \in \right)$ we partition X_t. That is, we write

$$X_t = X_t^{(m)} + X_t'^{(m)}$$

Where $X_t^{r(m)} = \sum_{i=m}^{\infty} \phi^j Z_{t-j}$, so that

$$0 \leq \bigvee_{t=1}^{n} \frac{X_{t-1}}{Z_t} - \bigvee_{t=1}^{n} \frac{X_{t-1}^{(m)}}{Z_t} \leq \bigvee_{t=1}^{n} \frac{X_{t-1}'^{(m)}}{Z_t}.$$

Define point processes

$$\xi_n^{(m)} = \sum_{t=1}^{n} \varepsilon_{\left(b_n^{-1} X_{t-1}'^{(m)}, Z_t\right)} \text{ and } \xi^{(m)} = \sum_{k=1}^{\infty} \sum_{i=m}^{\infty} \epsilon_{\left(\phi^i j_k, Z_{i,k}\right)}$$

where $\{j_k, k \geq 1\}$ and $\{Z_{i,k}, k \geq 1, i \geq 0\}$ have the distributional properties given in Theorem 2.2. Applying Theorem 2.2 with $c_i = \phi^i$ for $i \geq m$ and c_i equal to 0 otherwise, we obtain

$$\xi_n^{(m)} \xrightarrow{d} \xi^{(m)}.$$

Then letting for $x > 0$

$$\mathcal{R}_x = \left\{ (u, v) : u > 0, v > \theta, \text{and } \frac{u}{v} > x \right\},$$

we have

$$\lim_{n \to \infty} P\left(\bigvee_{t=1}^{n} b_n^{-1} X_{t-1}'^{(m)} / Z_t \leq x \right)$$
$$= \lim_{n \to \infty} P\left(\xi_n^{(m)} (\mathcal{R}_x) = 0 \right) = P\left(\xi^{(m)} (\mathcal{R}_x) = 0 \right).$$

Setting

$$V_k^{(m)} = \bigvee_{i=m}^{\infty} \frac{\phi^i}{Z_{i,k}} \text{ and } \zeta^{(m)} = \sum_{k=1}^{\infty} \epsilon_{\left(j_k, V_k^{(m)}\right)}$$

we have

$$P\left(\xi^{(m)} (\mathcal{R}_x) = 0 \right) = P\left(\bigvee_{k=1}^{\infty} j_k V_k^{(m)} \leq x \right)$$
$$= P\left(\zeta^{(m)} (\mathcal{R}_x) = 0 \right),$$

where

$$\mathcal{S}_x = \{(u,v) : u > 0, v > \theta, \text{and } uv > x\}.$$

Since $\xi^{(m)}$ is Poisson random measure with mean measure $\mu_m = v \times H_m$ where $V_1^{(m)} \sim H_m$ and

$$\mu_m(\mathcal{S}_x) = x^{-\beta} E\left(V_1^{(m)}\right)^{\beta},$$

we obtain

$$\lim_{n\to\infty} P\left(\bigvee_{t=1}^{n} b_n^{-1} X_{t-1}^{\prime(m)} / Z_t \le x\right) = \exp\left(-x^{-\beta} E\left(V_1^{(m)}\right)^{\beta}\right).$$

Next since

$$\left|\frac{1}{U_n^{(m)}} - \frac{1}{U_n}\right| \le \bigvee_{t=1}^{n} \frac{b_n^{-1} X_{t-1}^{\prime(m)}}{Z_t},$$

we have for large n that

$$P\left(\left|\frac{1}{U_n^{(m)}} - \frac{1}{U_n}\right| > \epsilon / M^2\right) \le P\left(\bigvee_{t=1}^{n} \frac{b_n^{-1} X_{t-1}^{\prime(m)}}{Z_t} > \epsilon / M^2\right)$$
$$\le 2\left(1 - \exp\left(-\epsilon^{-\beta} M^{2\beta} E\left(V_1^{(m)}\right)^{\beta}\right)\right)$$
$$\le 2\epsilon^{-\beta} M^{2\beta} E\left(V_1^{(m)}\right)^{\beta}.$$

Therefore, since $V_1^{(m)} \le \phi^m / \theta$,

$$\lim_{m\to\infty} \lim_{n\to\infty} P\left(\left|\frac{1}{U_n^{(m)}} - \frac{1}{U_n}\right| > \epsilon / M^2\right) = 0.$$

Next, note that from the limit law for the maximum obtained above, by replacing $X_{t-1}^{r(m)}$ with $X_{t-1}^{(m)}$ and by taking reciprocals, we derive the limit law for minimum,

$$\lim_{n\to\infty} P\left(U_n^{(m)} = \bigwedge_{t=1}^{n} \frac{Z_t}{b_n^{-1} X_{t-1}^{(m)}} \le x \right) \\ = 1 - \exp\left(-x^{\beta} E\tilde{W}_m^{-\beta}\right) \tag{2.19}$$

where $\tilde{W}_m$ has the distribution of $\wedge_{i=0}^{m-1} Z_{i,k} / \phi^t$. Thus, for any integer $m \ge 1$,

$$\lim_{n\to\infty} P\left(U_n^{(m)} \le x\right) \ge 1 - \exp\left(-(x/\theta)^{\beta}\right).$$

Thus for any $\in > 0$, we have for

M large enough that

$$\limsup_{n\to\infty} \sup_{m\ge 1} P\left(U_n^{(m)} > M\right) < \epsilon$$

completing the proof.

With Lemma 2.5 in hand we can now focus our attention to the limiting joint distribution of $(U_n^{(m)} > M) < \in$.

This will be accomplished by a blocking argument. To that end for a fixed positive integer k, let $r_n = \lfloor n/k \rfloor$ and define blocks for $i = 1, \cdots, \lfloor n/r_n \rfloor$ by

$$J_i = \left[(i-1)r_n + 1, \cdots, ir_n - q\right],$$
$$\text{and } J_i' = \left[ir_n - q + 1, \cdots, ir_n\right]$$

where q is a positive integer greater than m. Furthermore, let

$$J_0' = \left[r_n \lfloor n/r_n \rfloor + 1, \cdots, n\right].$$

Now we define the events

$$\chi_i = \left\{ \exists l \in J_i : \left(\frac{Z_l}{b_n^{-1} X_{l-1}^{(m)}} \right) \leq x \text{ or } a_n^{-1}\left(Z_l - \theta\right) \leq y \right\},$$

$i = 1, \cdots, l_n$

and

$$\chi_i' = \left\{ \exists l \in J_i' : \left(\frac{Z_l}{b_n^{-1} X_{l-1}^{(m)}} \right) \leq x \text{ or } a_n^{-1}\left(Z_l - \theta\right) \leq y \right\},$$

$i = 0, \cdots, l_n$

Where $l_n = \lfloor n / r_n \rfloor$. We begin by showing that the events x_i' are negligible.

Lemma 2.6: For any $x, y > 0$

$$\lim_{n \to \infty} P\left(\bigcup_{i=0}^{l_n} \chi_i' \right) = 0.$$

Proof: Observe that

$$nP\left(\frac{Z_l}{b_n^{-1} X_{l-1}^{(m)}} \leq x \right) \leq nP\left(\frac{\theta}{b_n^{-1} X_{l-1}^{(m)}} \leq x \right)$$
$$\sim x^{\beta} \theta^{-\beta} \sum_{i=0}^{m-1} \phi^{i\beta} \leq \frac{x^{\beta}}{\theta^{\beta}\left(1 - \phi^{\beta}\right)} \qquad (2.20)$$

and

$$nP\left(a_n^{-1}\left(Z_l-\theta\right)\leq y\right)\sim y^{\alpha}. \tag{2.21}$$

Thus for some constant c and any $n \geq 1$,

$$P\left(\bigcup_{i=0}^{l_n}\chi_i'\right)\leq ck/n$$

establishing the lemma.

Define events A_i and B_i by

$$A_i=\left\{\exists l\in J_i:\left(\frac{Z_l}{b_n^{-1}X_{l-1}^{(m)}}\right)\leq x\right\}$$

and $B_i=\left\{\exists l\in J_i:a_n^{-1}\left(Z_l-\theta\right)\leq y\right\}$.

The following result provides the asymptotic behavior of the probability of these events.

Lemma 2.7: For any $x>0, y>0$, we have as $k\to\infty$

$$\lim_{n\to\infty}P\left(A_i\right)\sim\frac{1}{k}x^{\beta}E\tilde{W}_m^{-\beta} \text{ and } \lim_{n\to\infty}P\left(B_i\right)\sim\frac{y^{\alpha}}{k}.$$

Proof: Since the events A_i are independent, we have

$$P\left(\bigcap_{i=1}^{k}\left(\frac{Z_l}{b_n^{-1}X_{l-1}^{(m)}}>x,\forall l\in J_i\right)\right)$$

$$=\left(P\left(\frac{Z_l}{b_n^{-1}X_{l-1}^{(m)}}>x,\forall l\in J_1\right)\right)^k.$$

Using Lemma 2.6 we have that

$$P\left(\bigwedge_{l=1}^{n}\frac{Z_l}{b_n^{-1}X_{l-1}^{(m)}} > x\right)$$

$$= P\left(\bigcap_{i=1}^{k}\left(\left(\frac{Z_l}{b_n^{-1}X_{l-1}^{(m)}}\right) > x, \forall l \in J_i\right)\right) + O\left(\frac{1}{n}\right)$$

$$= \left(P\left(\frac{Z_l}{b_n^{-1}X_{l-1}^{(m)}} > x, \forall l \in J_1\right)\right)^k + O\left(\frac{1}{n}\right).$$

From (2.19) we showed that

$$\lim_{n\to\infty} P\left(U_n^{(m)} = \bigwedge_{t=1}^{n}\frac{Z_t}{b_n^{-1}X_{t-1}^{(m)}} \le x\right)$$

$$= 1 - \exp\left(-x^{\beta} E\tilde{W}_m^{-\beta}\right).$$

Hence using this limit law on $\wedge_{t=1}^{n}(Z_t / b_n^{-1}X_{t-1}^{(m)})$, we obtain

$$\lim_{k\to\infty}\lim_{n\to\infty}\left(P\left[\frac{Z_l}{b_n^{-1}X_{l-1}^{(m)}} > x, \forall l \in J_1\right]\right)^k$$

$$= \exp\left(-x^{\beta} E\tilde{W}_m^{-\beta}\right).$$

Thus,

$$\lim_{n\to\infty} P[A_i] \sim \frac{x^{\beta}}{k} E\tilde{W}_m^{-\beta}, \text{as } k \to \infty. \tag{2.22}$$

Similarly using the result of Lemma 2.4, we obtain

$$\lim_{n\to\infty} P[B_i] \sim \frac{y^{\alpha}}{k}, \text{as } k \to \infty. \tag{2.23}$$

Hence the lemma holds.

Lemma 2.8: For some constant c

$$P(A_i \cap B_i) \le cP(A_i)P(B_i).$$

Remark. Since the cardinality of J_l depends on r_n which depends on k and the events A_i and B_i depend on n, $P(A_i)$ and $P(B_i)$ depend on k and n. The conclusion of this lemma provides that for all k and n, there is a constant dependent on no parameters for which the inequality stated there holds.

Proof: To calculate the intersection we define the following sets

$$K_1 = K_{1,i} = \{(l_1, l_2) : l_1, l_2 \in J_i \text{ and } l_2 \notin [l_1 - 1 - m, l_1]\}$$

and

$$K_2 = K_{2,i} = \{(l_1, l_2) : l_1, l_2 \in J_i \text{ and } l_2 \in [l_1 - 1 - m, l_1]\}.$$

Now for $l_1 - 1 - m \le l_2 \le l_1 - 1$ and n sufficiently large,

$$\left\{\left(\frac{Z_{l_1}}{b_n^{-1} X_{l_1-1}^{(m)}}\right) \le x\right\} \cap \{a_n^{-1}(Z_{l_2} - \theta) \le y\}$$

$$\subset \left\{\sum_{i \ne l_1 - l_2 - 1, 0 \le i \le m-1} \phi^i Z_{l_1-1-i} \ge b_n \frac{Z_{l_1}}{x} - \theta - 1\right\}$$

$$\cap \{Z_{l_2} \le \theta + a_n y\}.$$

It then follows from (2.20), (2.21), and independence that

$$P\left(\left\{\sum_{i\neq l_1-l_2-1,0\le i\le m-1}\phi^i Z_{l_1-1-i}\ge b_n\frac{Z_{l_1}}{x}-\theta-1\right\}\right.$$

$$\left.\cap\left\{Z_{l_2}\le\theta+a_n y\right\}\right)$$

$$=O\left(\frac{1}{n^2}\right).$$

If $l_1 = l_2 = l$, we have

$$P\left(\left\{\frac{Z_l}{b_n^{-1}X_{l-1}^{(m)}}\le x\right\}\cap\left\{a_n^{-1}\left(Z_l-\theta\right)\le y\right\}\right)$$

$$=\int_{\theta}^{\theta+a_n y}P\left(X_{l-1}^{(m)}\ge\frac{z}{b_n^{-1}x}\right)\mathrm{d}F(z)=O\left(\frac{1}{n^2}\right).$$

Therefore, for some constant c

$$P\left(\bigcup_{(l_1,l_2)\in K_2}\left\{\frac{Z_{l_1}}{b_n^{-1}X_{l_1-1}^{(m)}}\le x\right\}\cap\left\{a_n^{-1}\left(Z_{l_2}-\theta\right)\le y\right\}\right)$$

$$=c/nk.$$

In order to handle set K_1, observe from construction of the blocks J_l and set K_1 that if $(l_1,l_2)\in K_1$ then the events

$$\left\{\frac{Z_{l_1}}{b_n^{-1}X_{l_1-1}^{(m)}}\le x\right\}\text{ and }\left\{a_n^{-1}\left(Z_{l_2}-\theta\right)\le y\right\}$$

are independent. Thus, if we define $\{Z_i^{'}, i \in \mathbb{Z}\}$ as an independent copy of $\{Z_i^{'}, i \in \mathbb{Z}\}$, then

$$P\left(\bigcup_{(l_1,l_2)\in K_1}\left\{\frac{Z_{l_1}}{b_n^{-1}X_{l_1-1}^{(m)}}\le x\right\}\cap\left\{a_n^{-1}\left(Z_{l_2}-\theta\right)\le y\right\}\right)$$

$$=P\left(\bigcup_{(l_1,l_2)\in K_1}\left\{\frac{Z_{l_1}}{b_n^{-1}X_{l_1-1}^{(m)}}\le x\right\}\cap\left\{a_n^{-1}\left(Z_{l_2}'-\theta\right)\le y\right\}\right)$$

$$\le P\left(A_i\cap B_i'\right)=P\left(A_i\right)P\left(B_i\right)\le c/k^2$$

where $B_i^{'} = \{\exists l \in j_i : a_n^{-1}(Z_l^{'} - \theta) \le y\}$ and where we used Lemma 2.7 in the last step. Thus, we have that for some constant c

$$P\left(A_i\cap B_i\right)$$

$$=P\left(\bigcup_{(l_1,l_2)\in K_1}\left\{\frac{Z_{l_1}}{b_n^{-1}X_{l_1-1}^{(m)}}\le x\right\}\cap\left\{a_n^{-1}\left(Z_{l_2}-\theta\right)\le y\right\}\right.$$

$$\left.\cup\bigcup_{(l_1,l_2)\in K_2}\left\{\frac{Z_{l_1}}{b_n^{-1}X_{l_1-1}^{(m)}}\le x\right\}\cap\left\{a_n^{-1}\left(Z_{l_2}-\theta\right)\le y\right\}\right)$$

$$\le c/k^2=O\left(P\left(A_i\right)P\left(B_i\right)\right),$$

which completes the proof in view of Lemma 2.7.

Lemma 2.9: For any $x > 0, y > 0$,

$$\lim_{n\to\infty}P\left(\bigwedge_{t=1}^{n}\left(\frac{Z_t}{b_n^{-1}X_{t-1}^{(m)}}\right)>x, a_n^{-1}\bigwedge_{t=1}^{n}\left(Z_t-\theta\right)>y\right)$$

$$=\exp\left(-x^{\beta}E\tilde{W}_m^{-\beta}-y^{\alpha}\right).$$

Proof: First from Lemma 2.6, we have as $n \to \infty$ that

$$P\left(\bigwedge_{t=1}^{n}\left(\frac{Z_t}{b_n^{-1}X_{t-1}^{(m)}}\right) > x, a_n^{-1}\bigwedge_{t=1}^{n}(Z_t - \theta) > y\right)$$
$$= P\left(\bigcap_{i=1}^{k}\chi_i^c\right) + o(1).$$

Next by (2.22), (2.23), and Lemma 2.8 we obtain that as k tends to infinity

$$\lim_{n\to\infty} P(\chi_i) \sim \frac{1}{k}\left(x^{\beta} E\tilde{W}_m^{-\beta} + y^{\alpha}\right).$$

Therefore, we obtain

$$\lim_{k\to\infty}\lim_{n\to\infty} P\left(\bigcap_{i=1}^{k}\chi_i^c\right) = \lim_{k\to\infty}\left(1 - \frac{1}{k}\left(x^{\beta} E\tilde{W}_m^{-\beta} + y^{\alpha}\right)\right)^k$$
$$= \exp\left(-\left(x^{\beta} E\tilde{W}_m^{-\beta} + y^{\alpha}\right)\right).$$

Hence

$$\lim_{n\to\infty} P\left(\bigwedge_{t=1}^{n}\left(\frac{Z_t}{b_n^{-1}X_{t-1}^{(m)}}\right) > x, a_n^{-1}\bigwedge_{t=1}^{n}(Z_t - \theta) > y\right)$$
$$= \exp\left(-x^{\beta} E\tilde{W}_m^{-\beta} - y^{\alpha}\right).$$

Theorem 2.3: Let $\{X_t, t \geq 1\}$ denote the stationary AR(1) process such that the innovation distribution F satisfies $\bar{F}$ is $RV_{-\beta}$ at infinity and F is RV_{α} at θ.

If $\alpha > \beta$ then for any $x > 0, y > 0$ we have

$$\lim_{n\to\infty} P\left(b_n\left(\hat{\phi}_n - \phi\right) > x, a_n^{-1}\left(\hat{\theta}_n - \theta\right) > y\right)$$
$$= e^{-x^{\beta} EW^{-\beta} - y^{\alpha}},$$

Where $W = \wedge_{j=0}^{\infty} Z_j / \phi^j$.

Proof: Let us first observe that for $\epsilon > 0$

$$P\left(U_n^{(m)} > x + \epsilon, a_n^{-1} \bigwedge_{t=1}^{n} (Z_t - \theta) > y\right) - P\left(\left|U_n^{(m)} - U_n\right| > \epsilon\right)$$
$$\le P\left(U_n > x, a_n^{-1} \bigwedge_{t=1}^{n} (Z_t - \theta) > y\right)$$
$$\le P\left(U_n^{(m)} > x, a_n^{-1} \bigwedge_{t=1}^{n} (Z_t - \theta) > y\right).$$

Thus by Lemma 2.9 we obtain

$$\exp\left\{-(x+\epsilon)^{\beta} E\tilde{W}_m^{-\beta} - y^{\alpha}\right\}$$
$$- \limsup_{n\to\infty} P\left(\left|U_n - U_n^{(m)}\right| > \epsilon\right)$$
$$\le \liminf_{n\to\infty} P\left(U_n > x, a_n^{-1} \bigwedge_{t=1}^{n} (Z_t - \theta) > y\right)$$
$$\le \limsup_{n\to\infty} P\left(U_n > x, a_n^{-1} \bigwedge_{t=1}^{n} (Z_t - \theta) > y\right)$$
$$\le \exp\left(-x^{\beta} E\tilde{W}_m^{-\beta} - y^{\alpha}\right).$$

Letting m tend to infinity in the above and then $\in$ tend to 0, we obtain from Lemma 2.5 and $\lim_{m\to\infty} E\tilde{W}_m^{-\beta}$ that

$$\lim_{n\to\infty} P\left(U_n > x, a_n^{-1}\bigwedge_{t=1}^{n}(Z_t - \theta) > y\right)$$
$$= \exp\left(-x^{\beta} EW^{-\beta} - y^{\alpha}\right).$$

The theorem now follows from this and Corollary 2.2.

SIMULATION STUDY

In this section we assess the reliability of our extreme value estimation method through a simulation study. This included a comparison between our estimation procedure and that of three alternative estimation procedures for both the autocorrelation coefficient ϕ and the unknown location parameter θ under two different innovation distributions. Additionally, the degree of approximation for the empirical probabilities of $\hat{\phi}_{min}$ and $\hat{\theta}_{min}$ to its respective limiting distribution was reported.

To study the performance of the estimators

$\hat{\phi}_{min} = \wedge_{t=1}^{n} \frac{X_t}{X_{t-1}}$ and $\hat{\theta}_{min} = \wedge_{t\in I_n}(X_t - \hat{\phi}_{min} X_{t-1})$ respectively, we generated 5000 replications for the nonnegative time series $(X_0, X_1, \cdots, X_n)$ for two different sample sizes (500,1000), where $\{X_t\}$ is an AR(1) process satisfying the difference equation

$$X_t = \phi X_{t-1} + Z_t, \text{ for } (1 \le t \le n) \text{ and } Z_t \ge \theta.$$

The autoregressive parameter ϕ is taken to be in the range from 0 to 1 guaranteeing a nonnegative time series and the unknown location parameter θ is positive when the innovations Z_t are taken to be

$$F(z) = \begin{cases} c(z-\theta)^{\alpha}, \text{if } \theta < z < \theta+1, \\ 1-d(z-\theta)^{-\beta}, \text{if } \theta+1 < z < \infty. \end{cases}$$

For this innovation distribution let c and d be nonnegative constants such that c+d=1, then this distribution is regularly varying at both endpoints with index of regular variation $-\beta$ at infinity and index of regular variation α at θ. For this simulation study two distributions were considered:

i)$F_1, c = 0, d = 1$, ii)$F_2, c = 0.5, d = 0.5$.

Now observe in case i) the innovation distribution F_1 is a Pareto distribution with a regular varying tail distribution at ∞ with index of regular variation $-\beta$ and regular varying at $\theta+1$ with a fixed index $\alpha = 1$, whereas in case ii) the innovation distribution F_2 is regular varying at ∞ and θ with no restriction on α or β.

First we examine the simulation results for $\phi = 0.9$ under F_1 for each of the six different β values considered by computing 5000 estimates using

$$\hat{\phi}_{min} = \wedge_{t=1}^{n} \frac{X_t}{X_{t-1}}, \hat{\phi}_{max} = \frac{X_{t^*+1}}{X_{t^*}}, \hat{\phi}_{range} = \frac{X_{t^*+1} - X_{j^*}}{X_{t^*} - X_{j^*-1}}$$

where t* and j* provides the index of the maximal and minimal X_i respectively for $1 \le i \le n$, and

$$\hat{\phi}_{LS} = \begin{cases} \sum_{t=1}^{n-1} X_t X_{t+1} \Big/ \sum_{t=1}^{n} X_t^2, \text{if } 0 < \beta < 1 \\ \sum_{t=1}^{n} (X_t - \bar{X})(X_{t+1} - \bar{X}) \Big/ \sum_{t=1}^{n} (X_t - \bar{X})^2, \text{if } 1 < \beta < 3 \end{cases}$$

Where $\bar{X} = \sum_{t=1}^{n} X_t / n$. The means and standard deviations (written below in parentheses), of these estimates are reported in Table 1 along with the average length for a 95 percent empirical confidence intervals with exact coverage.

Since the main purpose of this section is to compare our estimator $\hat{\phi}_{min}$ to Bartlett and McCormick [5] estimator $\hat{\phi}_{max}$, McCormick and Mathew [2] estimator $\hat{\phi}_{range}$, and Davis and Resnick's [13] estimator $\hat{\phi}_{LS}$, the confidence intervals were directly constructed from the empirical distributions of

$n^{1/\beta}(\hat{\phi}_{min} - \phi), n^{1/\beta}(\hat{\phi}_{max} - \phi), n^{1/\beta}(\hat{\phi}_{range} - \phi)$ and $(n / \log n)^{1/\beta}(\hat{\phi}_{LS} - \phi)$ respectively.

To evaluate and compare the performance of four location estimators, six different scenarios for α and β are presented in Table 2 under F_2. When $\theta = 2$, 5000 estimates for each estimator;

$\hat{\theta}_{min} = \wedge_{t \in I_n}(X_t - \hat{\phi}_{min} X_{t-1})$, $\hat{\theta}_{range} = X_{j^*} - \hat{\phi}_{range} X_{j^*-1}$, $\hat{\theta}_{e1} = (1 - \hat{\phi}_{range}) \bigwedge_{t=1}^{n} X_t$, and

$\hat{\theta}_{e2} = (1 - \hat{\phi}_{LS}) \bigwedge_{t=1}^{n} X_t$ were obtained. The exponent p inside the index set

$I_n = \{t : 1 \le t \le n \text{ and } X_{t-1} \le (a_n b_n)^p\}$, was set to 0.9.

The means and standard deviations (written below in parentheses), of these estimates are reported in Table 2 along with the average length for a 95 percent empirical confidence intervals. For convenience, the empirical distributions of $n^{-1/\alpha}(\hat{\theta}_{min} - \theta)$, $n^{-1/\alpha}(\hat{\theta}_{range} - \theta)$,

$q_n(\hat{\theta}_{e1} - \theta - (1 - \hat{\phi}_{range})w_n)$, and

$q_n(\hat{\theta}_{e2} - \theta - (1 - \hat{\phi}_{LS})w_n)$ were respectively used, where the normalizing constants q_n and w_n are obtained through Equations (3.12)-(3.16) of McCormick and Mathew [2].

Remark. In the case that $0 < \beta < 1, \hat{\theta}_{e1}$, $\hat{\theta}_{e1}$ converge at

a faster rate than $\hat{\theta}_{e2}$ and in the case that $1 < \beta < 3, \hat{\theta}_{e2}$, $\hat{\theta}_{e2}$ converges at a faster rate than $\hat{\theta}_{e1}$. Lastly, since the McCormick and Mathew [2] paper has the restriction that $var(Z_1) < \infty$, only when $\alpha > 2$ can the estimators $\hat{\theta}_{e1}$ and $\hat{\theta}_{e2}$ be fairly compared, whereas only when $\alpha > \beta$ is our estimator applicable.

Now observe for the selected β values being considered, Table 1 shows that our estimator performs at least as well as the three other alternative estimators. This is particularly true under the heavier tail models, i.e. when $0 < \beta < 2$. In this regime our estimate shows little bias and the average lengths of the confidence intervals are smaller than the other three estimates, sometimes by a wide margin. In particular, when $\beta = 0.8$ and n = 1000 the 95% confidence interval average length for our method is 3.13, 5.78 and 23 times smaller than the three alternative estimators respectively. This is in part due to the use of one-sided confidence intervals since $\phi_{min} \geq \phi, \text{for all } t \geq 1$. Naturally, when $1 < \beta < 3$, Davis and Resnick least square estimator is more efficient than all three extreme value estimators. While our estimator $\hat{\phi}_{min}$ will always perform slightly better than the $\hat{\phi}_{max}$

estimator, Bartlett's and McCormick [5] estimator $\hat{\phi}_{max}$

Table 1: Comparison of estimators for $\phi = 0.9$ under F_1

β	n	$\hat{\phi}_{min}$	$\hat{\phi}_{max}$	$\hat{\phi}_{range}$	$\hat{\phi}_{LS}$	95% C.I. Avg. Length			
						Min est.	**Max est.**	**Range est.**	***LS* est.**
0.2	500	0.9000	0.9002	0.9002	0.8988	<0.0001	<0.0001	<0.0001	0.0288
		(<0.0001)	(0.0015)	(0.0016)	(0.0219)	-	-	-	
	1000	0.9000	0.9001	0.9002	0.8997	<0.0001	-0.0001	<0.0001	0.0091
		(<0.0001)	(0.0009)	(0.0009)	(0.0112)	-	-	-	
0.8	500	0.9004	0.9026	0.9032	0.9231	0.0046	0.0139	0.0228	0.0657
		(0.0016)	(0.0083)	(0.0083)	(0.0251)		-	-	
	1000	0.9002	0.9020	0.9014	0.9158	0.0023	0.0072	0.0133	0.0529
		(0.0008)	(0.0064)	(0.0064)	(0.0186)		-	-	
1.2	500	0.9049	0.9072	0.9067	0.8923	0.0271	0.0449	0.0548	0.0623
		(0.0078)	(0.0146)	(0.0139)	(0.0176)		-	-	
	1000	0.9029	0.9117	0.9047	0.8964	0.0187	0.0325	0.0389	0.0439
		(0.0056)	(0.0117)	(0.0109)	(0.0117)		-	-	
1.8	500	0.9191	0.9202	0.9136	0.8947	0.0562	0.0763	0.0846	0.0695
		(0.0622)	(0.0673)	(0.0612)	(0.0381)		-	-	
	1000	0.9113	0.9208	0.9116	0.8989	0.0501	0.0690	0.0743	0.0485
		(0.0313)	(0.0367)	(0.0384)	(0.0223)		-		

β	n	$\hat{\phi}_{min}$	$\hat{\phi}_{max}$	$\hat{\phi}_{range}$	$\hat{\phi}_{LS}$	95% C.I. Avg. Length			
						Min est.	Max est.	Range est.	*LS* est.
2.2	500	0.9236	0.9313	0.9176	0.8917	0.0670	0.0847	0.0944	0.0709
		(0.0909)	(0.1050)	(0.0244)	(0.0281)		-		
	1000	0.9207	0.9293	0.9151	0.8961	0.0623	0.0821	0.0874	0.0494
		(0.0412)	(0.0456)	(0.0217)	(0.0228)		-		
2.8	500	0.9475	0.9505	0.9205	0.8873	0.0767	0.0905	0.1050	0.0737
		(0.1378)	(0.1471)	(0.0723)	(0.0691)		-		
	1000	0.9437	0.9484	0.9187	0.8906	0.0735	0.0885	0.0956	0.0516
		(0.0834)	(0.1163)	(0.0317)	(0.0331)		-		

Table 2: Comparison of estimators for f = 2 under F_2

β	α	n	$\hat{\theta}_{min}$	$\hat{\theta}_{range}$	$\hat{\theta}_{e1}$	$\hat{\theta}_{e2}$	95% C.I. Avg. Length			
							Min est.	**Range est.**	**e1 (range) est.**	**e2 (LS) est.**
0.5	0.6	500	2.00	2.44	3.16	3.98	<0.0001	2.34	2.09	3.02
			(0.0087)	(0.4068)	(0.6775)	(0.7657)	-	-	-	
		1000	2.00	2.39	3.77	3.67	<0.0001	1.89	1.69	2.60
			(0.0002)	(0.3605)	(0.4381)	(0.4960)	-	-	-	
	1.6	500	1.94	2.68	3.34	4.10	0.0273	2.37	3.18	3.66
			(0.0879)	(0.3566)	(0.6656)	(0.7503)	-	-	-	
		1000	1.99	2.63	3.94	3.81	0.0152	1.87	2.23	2.79
			(0.0503)	(0.3199)	(0.4311)	(0.4780)	-	-	-	
	2.6	500	1.87	2.77	3.41	4.16	0.1422	2.19	2.85	2.38
			(0.4697)	(0.7553)	(0.6733)	(0.7632)	-	-	-	
		1000	1.89	2.73	3.99	3.87	0.0988	1.74	2.21	2.03
			(0.4045)	(0.3143)	(0.4276)	(0.4822)	-	-	-	
1.5	0.6	500	NA	1.72	2.35	2.15	NA	1.57	1.28	1.11
			(-)	(0.9410)	(0.5327)	(0.3706)	-	-		
		1000	NA	1.80	2.19	2.65	NA	1.20	1.25	1.88

β	α	n	$\hat{\theta}_{min}$	$\hat{\theta}_{range}$	$\hat{\theta}_{e1}$	$\hat{\theta}_{e2}$	95% C.I. Avg. Length			
							Min est.	Range est.	e1 (range) est.	e2 (LS) est.
			(-)	(0.9141)	(0.4381)	(0.2565)	-	-		
	1.6	500	2.80	1.95	2.54	2.26	1.72	2.23	2.18	1.94
			(0.3863)	(0.9569)	(0.5298)	(0.3815)	-		-	
		1000	2.38	2.03	2.37	2.84	1.28	2.10	1.78	1.31
			(0.2263)	(0.9308)	(0.4461)	(0.2701)	-	-	-	
	2.6	500	2.99	2.16	2.79	2.35	1.83	2.47	2.21	2.09
			(0.3604)	(0.9443)	(0.5271)	(0.3982)	-	-	-	
		1000	2.68	2.12	2.27	2.93	1.26	2.10	1.81	1.42
			(0.2956)	(0.9010)	(0.4398)	(0.2837)	-	-	-	

main advantage lies with its versatility to perform well for various non-negative time series, including but not restricted to higher order autoregressive models, along with ARMA models.

Table 2 reveals that our estimator for θ generally performs better than the three alternative estimators for $\alpha = 0.6, 1.6, 2.6$ when $0 < \beta < 1$. This is particularly true when comparing average confidence interval lengths. Although all three estimators $\hat{\theta}_{min}$, $\hat{\theta}_{range}$, and $\hat{\theta}_{el}$ converge to the true value of the parameter θ as n tends to infinity respectively, in this setting they may not compete asymptotically with, say, a conditional least square estimator $\hat{\theta}_{e2}$ when $\beta > 1$. Nonetheless for small sample sizes our simulation study favors $\hat{\theta}_{range}$ over the other three estimators. The difficulty for a least square estimate is that a small negative bias for the estimate of the autocorrelation parameter ϕ gives rise to a much larger positive bias in the estimate of $\hat{\theta}_{e2}$. While the affect is not as great, the positive bias found in our estimator $\hat{\phi}_{min}$ and the others for ϕ has a significant effect on the estimate for θ.

Figures 1-4 show a comparison be tween the probability that estimators $\hat{\phi}_{min}$, $\hat{\phi}_{max}$, $\hat{\phi}_{range}$, and $\hat{\phi}_{Ls}$ are within 0.01 of the true autocorrelation parameter value, respectively. With a sample size of 500, these figures plotted the sample fraction of estimates which fell within a bound of $\varepsilon = 0.01$ of the true value. Good performance with respect to this measure is reflected in curves near to 1.0 with diminishing good behavior as curves approach 0.0. When $0 < \beta < 1$, the figures seem to show that our estimator compared to the other three produced a higher fraction of precise estimates, especially compared to Davis and Resnick estimator.

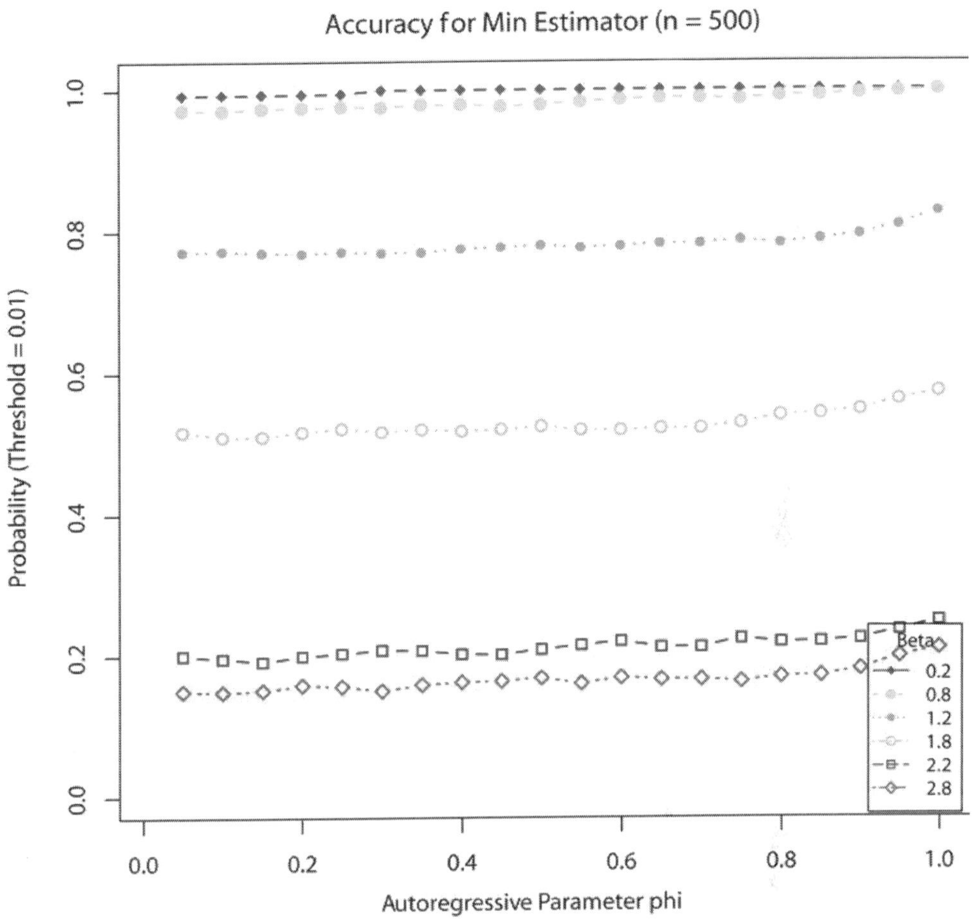

Figure 1: $P\left[|\hat{\phi}_{min} - \phi| < 0.01\right]$ for $RV_{-\beta}$.

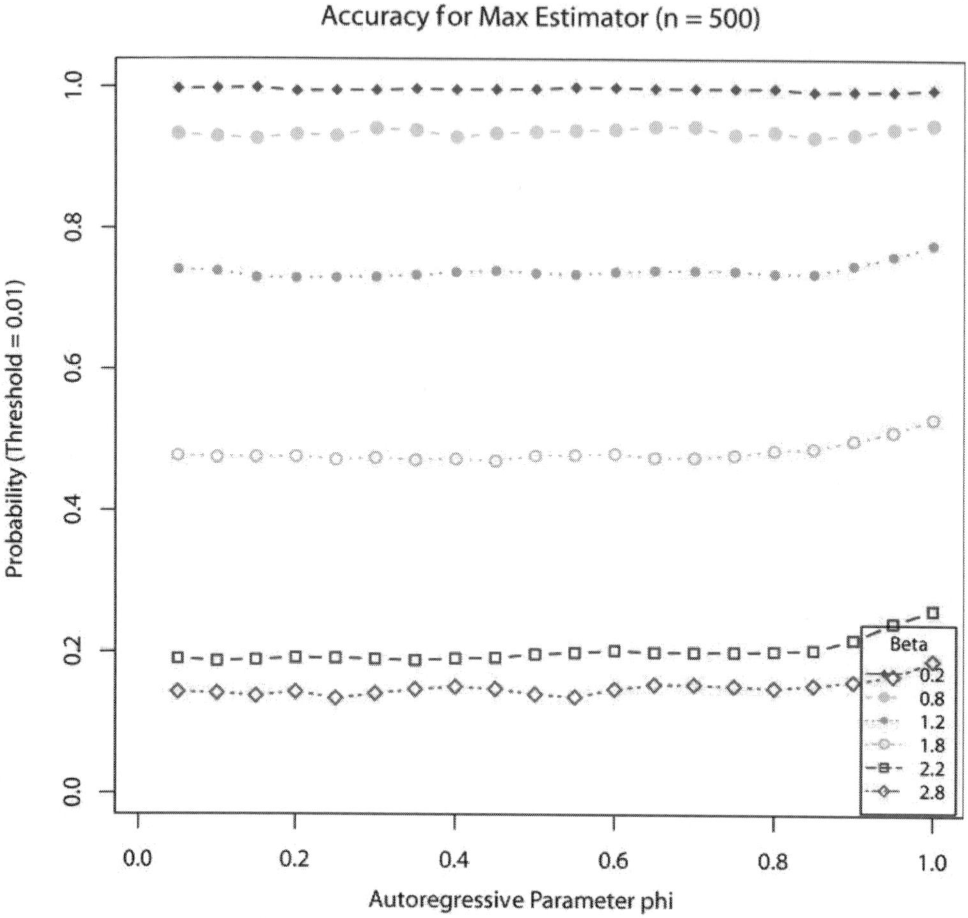

Figure 2: $P\left[\,|\hat{\phi}_{max} - \phi| < 0.01\right]$ for $RV_{-\beta}$.

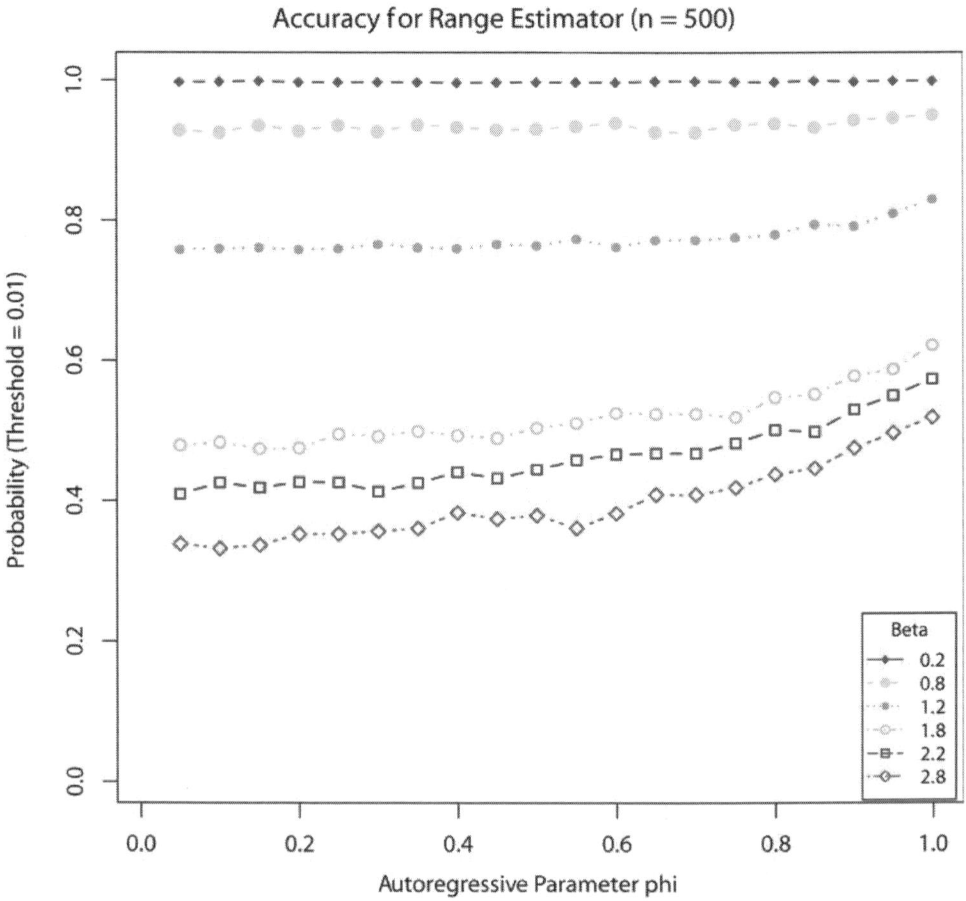

Figure 3: $P\left[|\hat{\phi}_{range} - \phi| < 0.01\right]$ for $RV_{-\beta}$.

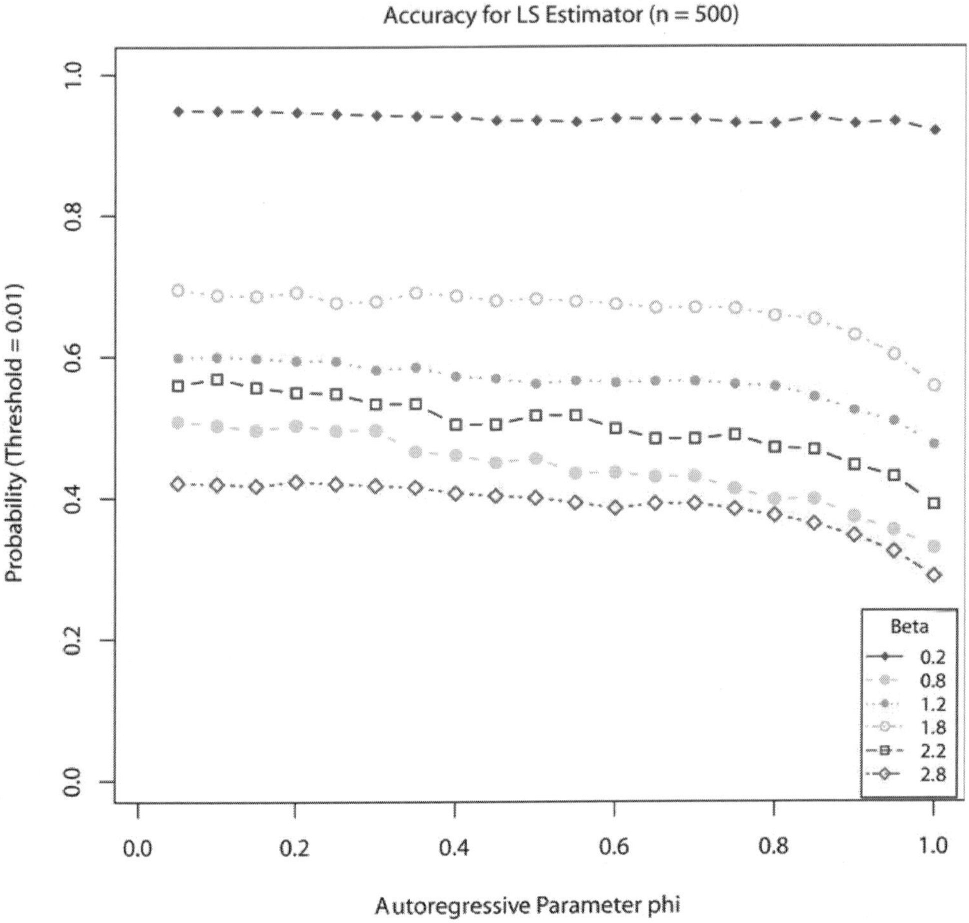

Figure 4: $P\left[|\hat{\phi}_{LS} - \phi| < 0.01\right]$ for $RV_{-\beta}$.

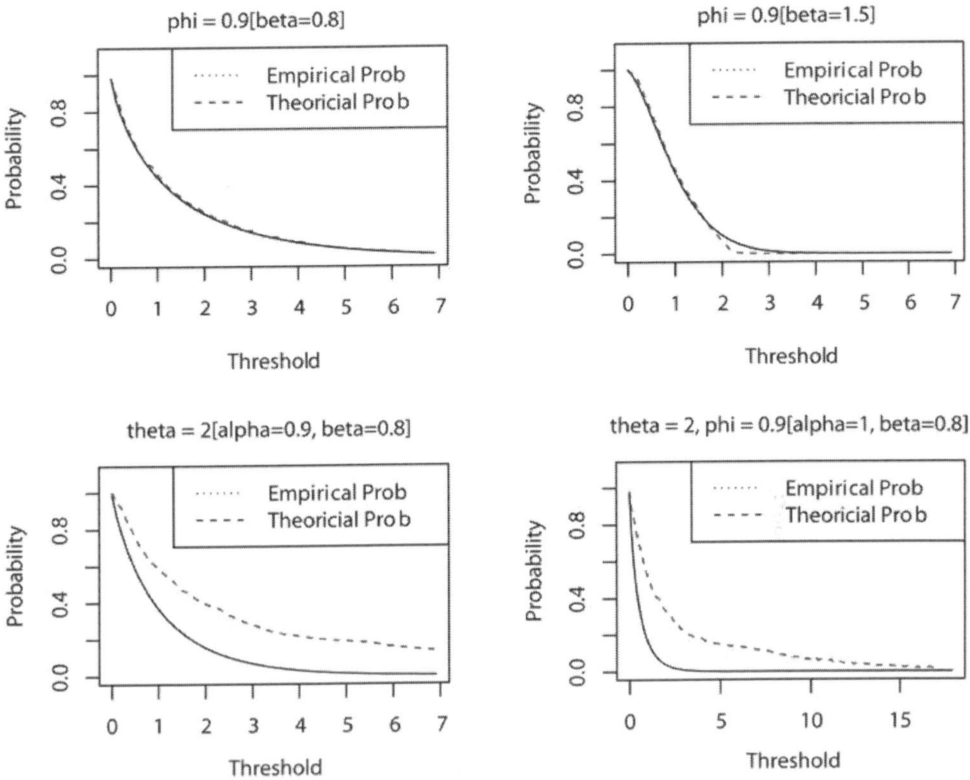

Figure 5: Empirical vs. theoretical probability.

When the regular variation index value is closer to 2, we see a higher fraction of the Davis-Resnick estimates showing better accuracy by this measure. The figures also indicate that McCormick and Mathew's range estimator produced a consistent high fraction of precise estimates when $\beta > 2$.

Lastly, we performed a Monte Carlo simulation to study the degree of approximation for the empirical probability $P\left[b_n(\hat{\phi}_{min} - \phi) > x\right]$, $P\left[a_n^{-1}(\hat{\phi}_{min} - \phi) > y\right]$, and $P\left[b_n(\hat{\phi}_{min} - \phi) > x, a_n^{-1}(\hat{\theta}_{min} - \theta) > x\right]$ to its limiting values $e^{-x^{\beta}EW^{-\beta}-y^{\alpha}}$, $e^{-y^{\alpha}}$, and $e^{-x^{\beta}EW^{-\beta}-y^{\alpha}}$ respectively. The empirical distributions were calculated from 5000 replications of the nonnegative time series $(X_0, X_1, \cdots X_n)$ for a sample size of 5000, where

$EW^{-\beta} = 1\sum_{i=0}^{M} \frac{\phi^{(\beta i(i+3)/2)}}{i+2}(1-\phi^{\beta(i+2)})$, and M was set to 500. Additionally, we restricted $\alpha > \beta$. The top two plots in Figure 5 below shows the performance when $Z_t \sim F_1$ and the autocorrelation coefficient ϕ is 0.9 for $\alpha = 1$ and β equal to 0.8, 1.5 respectively. Observe for $0 < x < 7$ that the empirical tail probability $b_n(\hat{\phi}_{min} - \phi) > x$ mirrors the theoretical probability quite nicely. The lower left plot in Figure 5 displays the asymptotic performance when $Z_t \sim F_2$ and the location parameter θ is 2 for $\alpha = 0.9, \beta = 0.8$. Notice that the convergence rate of the empirical probability to the theoretical probability is extremely slow. This is not surprising since on average our estimate falls more than 0.1 from the true value when $\beta = 0.8$. The lower right plot in Figure 5 displays the asymptotic performance when

$Z_t \sim F_1$ for the joint distribution of $(\hat{\phi}_{min}, \hat{\theta}_{min})$. Observe that this plot solidifies the asymptotic independence between $b_n(\hat{\phi}_{min} - \phi)$ and $a_n^{-1}(\hat{\theta}_{min} - \theta)$.

REFERENCES

1. P. J. Brockwell and R. A Davis, "Time Series: Theory and Methods," 2nd Edition, Springer, New York, 1987.
2. W. P. McCormick and G. Mathew, "Estimation for Nonnegative Autoregressive Processes with an Unknown Location Parameter," Journal of Time Series Analysis, Vol. 14, No. 1, 1993, pp. 71-92. doi:10.1111/j.1467-9892.1993.tb00130.x
3. A. E. Raftery, "Estimation Eficace Pour un Processus Autoregressif Exponentiel a Densite Discontinue," Publications de l'Institut de Statistique de l'Université de Paris, Vol. 25, No. 1, 1980, pp. 65-91.
4. R. A. Davis and W. P. McCormick, "Estimation for FirstOrder Autoregressive Processes with Positive or Bounded Innovations," Stochastic Processes and Their Applications, Vol. 31, No. 1, 1989, pp. 237-250. doi:10.1016/0304-4149(89)90090-2

5. A. Bartlett and W. P. McCormick, "Estimation for NonNegative Time Series with Heavy-Tail Innovations," Journal of Time Series Analysis, 2012. http://onlinelibrary.wiley.com/journal/10.1111/%28ISSN%291467-9892/earlyview
6. S. I. Resnick, "Point Processes, Regular Variation and Weak Convergence," Advances in Applied Probability, Vol. 18, No. 1, 1986, pp. 66-138. doi:10.2307/1427239
7. Jiří Anděl, "Non-Negative Autoregressive Processes," Journal of Time Series Analysis, Vol. 10, No. 1, 1989, pp. 1-11. doi:10.1111/j.1467-9892.1989.tb00011.x
8. Jiří Anděl, "Non-Negative Linear Processes," Applications of Mathematics, Vol. 36, No. 1, 1991, pp. 277-283.
9. S. Datta and W. P. McCormick, "Bootstrap Inference for a First-Order Autoregression with Positive Innovations," Journal of American Statistical Association, Vol. 90, No. 1, 1995, pp. 1289-1300. doi:10.1080/01621459.1995.10476633
10. S. I. Resnick, "Heavy-Tail Phenomena: Probabilistic and Statistical Modeling," Springer, New York, 2007.
11. R. A. Davis and S. Resnick, "Limit Theory for Moving Averages of Random Variables with Regularly Varying Tail Probabilities," Annals of Probability, Vol. 13, No. 1, 1985, pp. 179-195. doi:10.1214/aop/1176993074
12. O. Kallenberg, "Random Measures," Akademie-Verlag, Berlin, 1976.
13. R. A. Davis and S. Resnick, "Limit Theory for the Sample Covariance and Correlation Functions of Moving Averages," Annals of Statistics, Vol. 14, No. 1, 1986, pp. 533-558.doi:10.1214/aos/1176349937

CITATION

A. Bartlett and W. McCormick, "Estimation for Nonnegative First-Order Autoregressive Processes with an Unknown Location Parameter," Applied Mathematics, Vol. 3 No. 12A, 2012, pp. 2133-2147. doi:10.4236/am.2012.312A294.

Numerical Solutions For The Time-Dependent Emden-Fowler-Type Equations By B-Spline Method

Mohamed El-Gamel,
W. El-bashbashy, and
Atallah El-Shenawy
Department of Mathematical Sciences,
Faculty of Engineering, Mansoura
University, Mansoura, Egypt

10

ABSTRACT

A numerical method based on B-spline is developed to solve the time-dependent Emden-Fowler-type equations. We also present a reliable new algorithm based on B-spline to overcome the difficulty of the singular point at x=0. The error analysis of the method is described. Numerical results are given to illustrate the efficiency of the proposed method.

INTRODUCTION

In recent years, a lot of attentions have been devoted to the study of B-spline method to investigate various scientific models. The efficiency of the method has been formally proved by many researchers [1] - [7].

Spline functions have some attractive properties. Due to the being piecewise polynomial, they can be integrated and differentiated easily. Since they have compact support, numerical methods in which spline functions are used as a basis function lead to matrix systems including band matrices. Such systems have solution algorithms with low computational cost.

In this paper, we employ the B-spline method to solve the time dependent partial differential equations. For clarity, the method is presented for the heat equation

$$u_t = u_{xx} + \frac{r}{x}u_x + a\eta(x,t,u) + \mu(x,t), \quad 0 < x < 1, t > 0 \tag{1}$$

Subject to the boundary conditions

$$u_x(0,t) = 0, \; u(1,t) = \xi(t), \tag{2}$$

And the initial condition

$$u(x,0) = v(x). \tag{3}$$

Some forms of the above equations model several phenomena in mathematical physics and astrophysics such as the diffusion of heat perpendicular to the surface of parallel planes, theory of stellar structure, the thermal behavior of a spherical cloud of gas, isothermal gas sphere and theory of thermionic currents [8] -[10].

The solution of the time-dependent Emden-Fowler equation as well as a variety of linear and nonlinear singular IVPs in quantum mechanics and astrophysics is numerically challenging because of singularity behavior at the origin. The singularity behavior that occurs at the point x=0 is the main difficulty in the analysis of Equations (1)-(3).

The approximate analytical solutions to the Lane-Emden equations were presented by Shawagfeh [11] and Wazwaz [12] - [14] using the Adomian decomposition method (ADM). Very recently, Chowdhury and Hashim [15] applied the homotopy-perturbation method (HPM) and the variational iteration method (VIM) [16] to obtain approximate analytical solutions of the time-dependent Emden-Fowler-type equations.

The paper is organized as follows. In Section 2, we review some basic facts about the B-spline that are necessary for the formulation of the discrete linear and nonlinear system. In Section 3, the error analysis of the method is described. In Section 4, we formulate our B-spline collocation method to the solution of (1)-(3). In Section 5, numerical experiments are tested to demonstrate the viability of the proposed method.

SOME PROPERTIES OF B-SPLINE

B-splines are mathematically more sophisticated than other types of splines, so we start with a gentle introduction. We first use basic assumptions to derive the expressions for the cubic uniform B-splines directly and without mentioning knots. We then show how to extend the derivations to uniform B-splines of any order. Following this, we discuss a different, recursive formulation of the weight functions of the uniform B-splines.

The third-degree B-spline is used to construct numerical solutions to a given problem (1). A detailed description of third-degree B-spline functions can be found in [13] [17]. The third-degree B-splines are defined as

$$B_{i-1}(x) = B_0\left(x-(i-1)h\right), \quad i = 2,3,\mathrm{L} \tag{4}$$

$$B_0(x) = \frac{1}{6h^3}\begin{cases} x^3 & \text{if; } 0 \le x < h \\ -3x^3+12hx^2-12h^2x+4h^3, & \text{if; } h \le x < 2h \\ 3x^3-24hx^2+60h^2x-44h^3, & \text{if; } 2h \le x < 3h \\ -x^3+12hx^2-48h^2x+64h^3 & \text{if; } 3h \le x < 4h \end{cases} \tag{5}$$

We can simply check that each of the function $B_i(x)$ is twice continuously differentiable on the entire real line. Also,

$$B_i(x_j) = \begin{cases} 4, & \text{if; } i = j \\ 1, & \text{if; } i-j = \pm 1 \\ 0, & \text{if. } i-j = \pm 2 \end{cases} \tag{6}$$

Similarly, we can show that:

$$B_i'(x_j) = \begin{cases} 0, & \text{if; } i = j \\ \pm\frac{3}{h}, & \text{if; } i - j = \pm 1 \\ 0, & \text{if. } i - j = \pm 2 \end{cases} \tag{7}$$

And

$$B_i''(x_j) = \begin{cases} -\frac{12}{h^2}, & \text{if; } i = j \\ \frac{6}{h^2}, & \text{if; } i - j = \pm 1 \\ 0, & \text{if. } i - j = \pm 2 \end{cases} \tag{8}$$

ERROR ANALYSIS

Governing Equation

Since x=0 is singular point of Equation (1), we first modify Equation (1) at x=0 By L'Hospital rule, the boundary value problem (1) is transform into

$$u_t = p(x)u_{xx} + q(x)u_x + a\eta(x,t,u) + \mu(x,t), \tag{9}$$

Where

$$p(x) = \begin{cases} r+1, & x = 0; \\ 1, & x \neq 0 \end{cases} \qquad q(x) = \begin{cases} 0, & x = 0; \\ \frac{r}{x}, & x \neq 0 \end{cases}$$

Let $\eta(x,t,u)=h(x,t)u$, we consider the linear non-homogeneous Emden-Fowler differential Equation (9), difference schemes for this problem considered as following:

$$p(x)\frac{d^2u_{i+1}}{dx^2}+q(x)\frac{du_{i+1}}{dx}+ah(x,t)u_{i+1}+\mu(x,t)=\frac{u_{i+1}-u_i}{\Delta t} \tag{10}$$

Or

$$(\Delta t)p(x)u_{i+1}^{(2)}+\Delta tq(x)u_{i+1}^{(1)}+[a(\Delta t)h(x,t)-1]u_{i+1}=-u_i-(\Delta t)\mu(x,t) \tag{11}$$

Suppose S(x) is the cubic B-spline interpolating u (x_k, ti_{+1}) at the $(i+1)^{th}$ time level then

$$S(x)=\sum_{k=-3}^{n-1}C_kB(x) \tag{12}$$

Theorem 3.1. Let the collocation approximation S(x) to the solution u(x) of the boundary value problem (1) is (12) then the following relation is hold

$$S'(x)=u_x(x_i)-\frac{h^4}{180}u_{xxxxx}(x_i)+O(h^6)$$
$$S''(x)=u_{xx}(x_i)-\frac{h^2}{12}u_{xxxx}(x_i)+\frac{h^4}{360}u_{xxxxxx}+O(h^6) \tag{13}$$

Truncation Error of Time Dependent Emden-Fowler Equation

Theorem 3.2. By using the combination of the finite difference approximation and cubic B-spline method, the truncation error is $O(\Delta t,h^2)$.

Proof: Consider the Equation (10) when apply finite difference

$$\frac{1}{\Delta t}(u_{i+1}-u_i)=p(x)u_{xx}+q(x)u_x+ah(x,t)u+\mu(x,t)$$

Apply Taylor expansion for L.H.S

$$\frac{1}{\Delta t}\left(u_i + \Delta t u_t + \frac{(\Delta t)^2}{2!}u_{tt} + \frac{(\Delta t)^3}{3!}u_{ttt} + \mathrm{L} - u_i\right) = p(x)u_{xx} + q(x)u_x + ah(x,t)u + \mu(x,t)$$

$$u_t + \frac{\Delta t}{2!}u_{tt} + \frac{(\Delta t)^2}{3!}u_{ttt} + \mathrm{L} = p(x)\left(u_{xx} - \frac{1}{12}h^2 u_{xxxx} + \frac{1}{360}h^4 u_{xxxxxx} + O(h^6)\right)$$
$$+ q(x)\left(u_x - \frac{1}{180}h^4 u_{xxxx} + O(h^6)\right) + ah(x,t)u + \mu(x,t)$$

If

$$p(x)u_{xx} + q(x)u_x + ah(x,t)u + \mu(x,t) = \lambda(x,t)$$

We can calculate the truncation error which is defined as $e_i = u_t - \lambda$ (xx,t) as:

$$e = \frac{\Delta t}{2!}u_{tt} + \frac{(\Delta t)^2}{3!}u_{ttt} + \mathrm{L} + \left(\frac{h^2}{12} - \frac{h^4}{180}\right)u_{xxxx} - O(h^6)$$

B-SPLINE SOLUTIONS FOR TIME-DEPENDENT EMDEN-FOWLER

In this section, we shall introduce a reliable algorithm based on B-spline method to handle singular initial value problems (IVPs) in a realistic and efficient way considering Emden Fowler equation as a model problem.

Linear Time-Dependent Emden-Fowler

Let

$$u(x) = \sum_{j=-3}^{n-1} C_j B_j(x) \tag{14}$$

Be an approximate solution of Equation (1), where C_i are unknown real coefficients and $B_j(x)$ are thirddegree B-spline functions. Let $x_0, x_1, \mathrm{L}, x_n$ be n+2 $n+1$

Grid points in the interval [α, b] so that

$$x_i = a + ih,\ i = 0,1,\ldots,n,\quad x_0 = a,\ x_n = b,\ h = \frac{b-a}{n}$$

We can deduce the following

$$u^{(1)}(x) = \sum_{j=-3}^{n-1} C_j B_j'(x)$$

$$u^{(2)}(x) = \sum_{j=-3}^{n-1} C_j B_j''(x) \qquad (15)$$

Theorem 4.1. If the assumed approximate solution of the problem (10) is (14), then the discrete collocation system for the determination of the unknown coefficients $\{C_j\}_{j=-3}^{n-1}$ is given by

$$(\Delta t)p(x_i)\sum_{j=-3}^{n-1} C_j B_j''(x_i) + \Delta t q(x_i)\sum_{j=-3}^{n-1} C_j B_j'(x_i)$$

$$+\left[a(\Delta t)h(x_i,t)-1\right]\sum_{j=-3}^{n-1} C_j B_j(x_i) = -u_i - (\Delta t)\mu(x_i,t),\quad 0 \le i \le n. \qquad (16)$$

Proof: If we replace each term of (10) with its corresponding approximation given by (14) and (15) and substituting $x=x_k$ and applying the collocation to it.

The boundary conditions (3) can be written as

$$\sum_{j=-3}^{n-1} C_j B_j'(0) = 0$$

$$\sum_{j=-3}^{n-1} C_j B_j(1) = \xi(t) \qquad (17)$$

Then use them with the system of Equation (16), which can be written in the matrix form

$$\boldsymbol{AC} = \boldsymbol{F} \tag{18}$$

Where

$$\boldsymbol{C} = \begin{pmatrix} C_{n-1} \\ C_{n-2} \\ \mathrm{M} \\ C_{-2} \\ C_{-3} \end{pmatrix}, \quad \boldsymbol{F} = \begin{pmatrix} 0 \\ \mathrm{M} \\ \xi(t) \end{pmatrix}$$

Also the matrix A can be written as

$$\boldsymbol{A} = \begin{pmatrix} \frac{-3}{h} & 0 & \frac{3}{h} & \mathrm{L} & 0 & 0 \\ \alpha_1(x_0) & \alpha_2(x_0) & \alpha_3(x_0) & \mathrm{L} & 0 & 0 \\ 0 & \alpha_1(x_1) & \alpha_2(x_1) & & 0 & 0 \\ \mathrm{M} & \mathrm{M} & \mathrm{M} & \mathrm{M} & \mathrm{M} & \mathrm{M} \\ 0 & 0 & \mathrm{L} & & \alpha_3(x_{n-1}) & 0 \\ 0 & 0 & \mathrm{L} & & \alpha_2(x_n) & \alpha_3(x_n) \\ 0 & 0 & \mathrm{L} & & 4 & 1 \end{pmatrix}$$

Where

$$\alpha_1(x_i) = p(x_i)\left(\frac{6}{h^2}\right) - q(x_i)\left(\frac{3}{h}\right) + \left[a(\Delta t)\eta(x_i,t) - 1\right]$$

$$\alpha_2(x_i) = p(x_i)\left(-\frac{12}{h^2}\right) + \left[a(\Delta t)\eta(x_i,t) - 1\right]\cdot 4$$

$$\alpha_3(x_i) = p(x_i)\left(\frac{6}{h^2}\right) + q(x_i)\left(\frac{3}{h}\right) + \left[a(\Delta t)\eta(x_i,t) - 1\right]$$

It is easily seen that the matrix A is strictly diagonally dominant and hence nonsingular. Now we have a linear system of n+3 equations for the n=3 unknown coefficients, namely, $\{C_j\}_{j=-3}^{n-1}$. We can obtain the coefficient of the approximate solution by solving this linear system by Q-R method.

Non-Linear Time-Dependent Emden-Fowler

Let $\eta(x, t, u) = h(x, t)\, g(u)$, we consider the nonlinear non-homogeneous Emden-Fowler differential equation

$$p(x)u_{xx} + q(x)u_x + ah(x,t)g(u) + \mu(x,t) = u_t \tag{19}$$

We can use finite difference method

$$p(x)\frac{d^2 u_{i+1}}{dx^2} + q(x)\frac{du_{i+1}}{dx} + ah(x,t)g(u_{i+1}) + \mu(x,t) = \frac{u_{i+1} - u_i}{\Delta t} \tag{20}$$

Theorem 4.2. If the assumed approximate solution of the problem (20) is (14), then the discrete collocation system for the determination of the unknown coefficients is given by

$$p(x)\sum_{j=-3}^{n-1} C_j B_j''(x_k) + q(x)\sum_{j=-3}^{n-1} C_j B_j'(x_k) + ah(x_k, t_{i+1})\, g\left(\sum_{j=-3}^{n-1} C_j B_j(x_k)\right)$$

$$+\mu(x_k, t_{i+1}) = \frac{\sum_{j=-3}^{n-1} C_j B_j(x_k) - u_i}{\Delta t}, \quad k = 0,1,\mathrm{L}\ ,n. \tag{21}$$

Then the boundary condition (17) with the system of Equation (21). Now we have a nonlinear system of n=3 equations in the n+3 unknown coefficients. We can obtain the coefficients of the approximate solution by solving this nonlinear system by Newton's method.

EXAMPLES AND COMPARISONS

In this section, we will present three of our numerical results of time dependent problems using the method outlined in the previous section. The performance of the B-spline method is measured by the maximum absolute error ε_k which is defined as

$$\varepsilon_k = \left| u_{\text{exact}} - u_{\text{B-spline}} \right|$$

Example 1: [18] Consider the problem

$$u_t = u_{xx} + \frac{2}{x} u_x - \left(6 + 4x^2 - \cos t\right) u, \quad 0 < x < 1, t > 0$$

Subject to boundary conditions

$$u_x\left(0, t\right) = 0, \quad u\left(1, t\right) = \mathrm{e}^{1+\sin t},$$

And the initial condition

$$u\left(x, 0\right) = \mathrm{e}^{x^2}.$$

Whose exact solution is?

$$u\left(x, t\right) = \mathrm{e}^{x^2 + \sin t}.$$

The maximum absolute errors at different n and different time t with $\Delta t = 0.001$ is given inTable 1.

Example 2: [18] Consider the linear problem

$$u_t + \left(6 - 5x^2 - 4x^4\right) = u_{xx} + \frac{2}{x} u_x - \left(5 + 4x^2\right) u,$$

Subject to boundary conditions

$$u_x\left(0, t\right) = 0, \quad u\left(1, t\right) = 1 + \mathrm{e}^{1+t},$$

And the initial condition

$$u\left(x, 0\right) = x^2 + \mathrm{e}^{x^2}.$$

Whose exact solution is?

Table 1: Maximum absolute errors for Example 1

n	T = 0.1	T = 0.5	T = 1.0
10	0.002756	0.00493	0.007185
20	6.556E−04	0.00119	0.001819
30	2.693E−04	5.124E−04	8.373E−04
40	1.348E−04	2.742E−04	4.947E−04
50	7.304E−05	1.643E−04	3.364E−04
60	3.999E−05	1.049E−04	2.506E−04
70	2.074E−05	6.951E−05	1.990E−04
80	1.261E−05	4.683E−04	1.657E−04

$$u(x,t) = e^{x^2+t} + x^2.$$

The maximum absolute errors at different n and different time t with $\Delta t = 0.001$ is given inTable 2.

Example 3: [18] Now we turn to a singular nonlinear problem

$$u_t = u_{xx} + \frac{5}{x}u_x - \left(24t + 16t^2x^2\right)e^u - 2x^2e^{u/2},$$

Subject to boundary conditions

$$u_x(0,t) = 0, \quad u(1,t) = -2\ln(1+t),$$

And the initial condition

$$u(x,0) = 0.$$

Whose exact solution is?

$$u(x,t) = -2\ln\left(1+tx^2\right).$$

The maximum absolute errors at different n and different time t with $\Delta t = 0.001$ is given inTable 3.

CONCLUSIONS

We have presented a B-spline collocation method for the solution of time-dependent Emden Fowler type Equations (1)-(3). This method produces a spline function which is useful to obtain the solution at any point of the interval, whereas the finite difference method gives the solution only at selected nodal points. The numerical results given in tables show that the present method approximates the exact solution very well.

This present analysis exhibits the reliable applicability of B-spline method to solve linear and nonlinear problems with singular feature. In this work, we demonstrate that this method can be well suited to attain solution to the type of examined equations, linear and nonlinear as well. The difficulty in this type of equations

Table 2: Maximum absolute errors for Example 2

n	T = 0.1	T = 0.5	T = 1.0
10	0.002756	0.00498	0.00821
20	6.528E−04	0.00117	0.00192
30	2.664E−04	4.729E−04	7.799E−04
40	1.319E−04	2.309E−04	3.807E−04
50	7.021E−05	1.203E−04	1.984E−04
60	3.727E−05	6.203E−05	1.022E−04
70	1.823E−05	2.906E−05	4.791E−05
80	1.541E−05	2.431E−05	4.008E−05

Table 3: Maximum absolute errors for Example 3

n	T = 0.01	T = 0.1	T = 0.2
10	6.597E−05	5.805E−04	0.001176
20	1.494E−05	1.340E−04	2.807E−04
30	6.481E−06	5.202E−05	1.148E−04

Due to the existence of singular point at x=0, is overcome here. Finally, we conclude that the B-spline method is a promising tool for both linear and nonlinear singular time-dependent Emden-Fowler-type equations.

REFERENCES

1. Ali, A.H.A., Gardner, G.A. and Gardner, L.R.T. (1992) A Collocation Solution for Burgers? Equation Using Cubic B-Spline Finite Elements. Computer Methods in Applied Mechanics and Engineering, 100, 325-337. http://dx.doi.org/10.1016/0045-7825(92)90088-2
2. Aziz, T., Khan, A. and Rashidinia, J. (2005) Spline Methods for the Solution of Fourth-Order Parabolic Partial Differential Equations. Applied Mathematics and Computation, 167, 153-166. http://dx.doi.org/10.1016/j.amc.2004.06.095
3. Caglar, N. and Caglar, H. (2006) B-Spline Solution of Singular Boundary Value Problems. Applied Mathematics and Computation, 182, 1509-1513.http://dx.doi.org/10.1016/j.amc.2006.05.035
4. Caglar, H., Ozer, M. and Caglar, N. (2008) The Numerical Solution of the One-Dimensional Heat Equation by Using Third Degree B-Spline Functions. Chaos, Solitons & Fractals, 38, 1197-1201. http://dx.doi.org/10.1016/j.chaos.2007.01.056
5. Fazal-i-Haq, Siraj-ul-Islam and Tirmizi, I.A. (2010) A Numerical Technique for Solution of the MRLW Equation Using Quartic B-Splines. Applied Mathematical Modelling, 34, 4151-4160. http://dx.doi.org/10.1016/j.apm.2010.04.012
6. Lakestani, M. and Dehghan, M. (2012) Numerical Solutions of the Generalized Kuramoto-Sivashinsky Equation Using B-Spline Functions. Applied Mathematical Modelling, 36, 605-617. http://dx.doi.org/10.1016/j.apm.2011.07.028
7. Mittal, R.C. and Jain, R.K. (2011) B-Splines Methods with Redefined Basis Functions for Solving Fourth Order Parabolic Partial Differential Equations. Applied Mathematics and Computation, 217, 9741-9755. http://dx.doi.org/10.1016/j.amc.2011.04.061
8. Chandrasekhar, S. (1967) Introduction to the Study of Stellar Structure. Dover, New York.

9. Davis, H.T. (1962) Introduction to Nonlinear Differential and Integral Equations. Dover, New York.
10. Richardson, O.U. (1921) the Emission of Electricity from Hot Bodies. Logmans Green, London.
11. Shawagfeh, N.T. (1993) Nonperturbative Approximate Solution for Lane-Emden Equation. Journal of Mathematical Physics, 34, 4364-4369. http://dx.doi.org/10.1063/1.530005
12. Wazwaz, A. (2005) Analytical Solution for the Time-Dependent Emden-Fowler Type of Equations by Adomian Decomposition Method. Applied Mathematics and Computation, 166, 638-665. http://dx.doi.org/10.1016/j.amc.2004.06.058
13. Wazwaz, A.M. (2005) Adomian Decomposition Method for a Reliable Treatment of the Emden-Fowler Equation. Applied Mathematics and Computation, 161, 543-560.http://dx.doi.org/10.1016/j.amc.2003.12.048
14. Wazwaz, A. (2001) A New Algorithm for Solving Differential Equations of Lane?—Emden Type. Applied Mathematics and Computation, 118, 287-310. http://dx.doi.org/10.1016/S0096-3003(99)00223-4
15. Chowdhury, M.S.H. and Hashim, I. (2007) Solutions of Time-Dependent Emden? Fowler Type Equations by Homotopy-Perturbation Method. Physics Letters A, 368, 305-313.http://dx.doi.org/10.1016/j.physleta.2007.04.020
16. Wazwaz, A. (2011) The Variational Iteration Method for Solving Nonlinear Singular Boundary Value Problems Arising in Various Physical Models. Communications in Nonlinear Science and Numerical Simulation, 16, 3881-3886.http://dx.doi.org/10.1016/j.cnsns.2011.02.026
17. de Boor, C. (1978) A Practical Guide to Splines. Springer Verlag, New York.http://dx.doi.org/10.1007/978-1-4612-6333-3
18. Matinfar, M. and Mahdavi, M. (2012) Approximate Analytical Solutions for the Time-Dependent Emden-Fowler-Type Equations by Variational Homotopy Perturbation Method. WJMS, 8, 58-65

CITATION

El-Gamel, M. , El-bashbashy, W. and El-Shenawy, A. (2014) Numerical Solutions for the Time-Dependent Emden-Fowler-Type Equations by B-Spline Method. Applied Mathematics, 5, 593-600. Doi: 10.4236/am.2014.54056.

Modeling of Surface Waves in a Fluid Saturated Poro-Elastic Medium under Initial Stress Using Time-Space Domain Higher Order Finite Difference Method

11

Anjana P. Ghorai[1] *and R. Tiwary*[2]

[1]Department of Applied Mathematics, Birla Institute of Technology, Ranchi, India

[2]Department of Mathematics, RVS College of Engineering & Technology, Jamshedpur, India

ABSTRACT

In this present context, mathematical modeling of the propagation of surface waves in a fluid saturated poro-elastic medium under the influence of initial stress has been considered using time dependent higher order finite difference method (FDM). We have proved that the accuracy of this finite-difference scheme is 2M when we use 2nd order time domain finite-difference and 2M-th order space domain finite-difference. It also has been shown that the dispersion curves of Love waves are less dispersed for higher order FDM than of lower order FDM. The effect of initial stress, porosity and anisotropy of the layer in the propagation of Love waves has been studied here. The numerical results have been shown graphically. As a particular case, the phase velocity in a non porous elastic solid layer derived in this paper is in perfect agreement with that of Liu et al. (2009).

INTRODUCTION

The simulation of surface waves propagating in a fluid saturated poro-elastic media is of great importance to seismologists due to its possible applications in geophysical prospecting, reservoir engineering and survey techniques for understanding the cause and estimation of damage due to natural and manmade hazards. Also the difficulty in exploring natural resources is gradually increasing as the nature of the reservoir is more complex and heterogeneous than it was assumed in past and the characterization of the subsurface materials as fluid saturated porous media is more realistic. It is also more accurate to consider soil as two phase composite materials, granular solid and pore fluid. The size of pores is assumed to be small and macroscopically speaking, their average distribution is uniform.

Since poro-elastic theory was developed by Biot [1-3] many efforts have been made in using experimental and numerical methods to characterize elastic wave propagation in porous, liquid-saturated solids. Quite a good amount of information about the effect of initial stress on propagation of surface waves is available in the literature of many authors namely, A. M. Abd-Alla [4], S. Gupta [5], Shishir Gupta [6]. In these papers they have followed the conventional analytical method to solve these problems. But recently the finite difference method appears as an important tool for numerical simulations of partial differential equations and has been used widely in simulating elastic waves. Due to easy and straightforward approach, robustness, requiring small memory and computation time, the Finite Difference Methods are the most popular methods for seismic modeling. Also wave equation modeling in the time domain is popular because of its easy implementation and accuracy, compared to frequency domain modeling. In time-domain modeling, higher-order differencing operators are used to reduce the required spatial sampling. Virieux [7, 8] have used velocity-stress finite difference method for the propagation of P-SV wave and SH wave in heterogeneous media. To improve the accuracy and stability of finite difference scheme, many authors have used and developed different types of difference schemes. Levander [9] applied 4th order approximation in space to the P-SV scheme. Hayashi and burns [10] developed finite difference scheme with variable grids. R. W. Graves [11] discussed the simulation of seismic wave propagation in 3D elastic media using staggered-grid finite difference method. Kristek et al. [12] considered seismic wave propagation in viscoelastic media using 3D forth order staggeredgrid finite difference scheme. Saenger et al. [13]

have considered the rotated staggered grid finite difference method. Bohlen et al. [14] discussed the accuracy of heterogeneous staggered-grid finite-difference modeling. Kristek et al. [15] discussed on the accuracy of the finite-difference schemes for the one dimensional elastic problem. Tessmer [16] discussed Seismic finite difference modeling with spatially variable time steps. Finkelstein et al. [17] developed finite difference time domain dispersion reduction schemes. Y. Liu et al. [18,19] discussed advanced and truncated finite difference method for seismic modeling and Y. Liu et al. [20] employed a plane wave theory and the Taylor series expansion of dispersion relation to derive the FD coefficients in the joint time-space domain for the scalar wave equation with second-order spatial derivatives. They demonstrated that the method has greater accuracy and better stability than the conventional method. Dublain [21] has demonstrated the advantages of the higher order difference scheme. Lie et al. [22] designed spatial finite difference stencils on a time-space domain to simulate wave propagation in acoustic vertically transversely isotropic medium. Two-dimensional dispersion analysis and numerical modeling demonstrate that this stencil has greater precision than one used in a conventional F. D. Lie et al. [23] discussed finite difference numerical modeling in two phase anisotropic media with even order accuracy. Zhu et al. [24] developed finite difference modeling of the seismic response of fluid saturated, porous, elastic solid using Biot theory. Lie et al. [25] developed a time-space domain dispersion-relationbased staggered-grid finite-difference schemes for modeling the scalar wave equation. Also, the new stencils can adopt a larger time step. Dispersion analysis and numerical modeling results demonstrate that the new stencils have greater accuracy and can effectively suppress the dispersion and retain the waveform.

In this paper, following [22,25] we have modeled the Love waves in a fluid saturated poro-elastic media under the influence of initial stress using time-space domain higher order finite difference method. The dispersion equation has been obtained and the presence of the porosity parameter, the non-dimensional anisotropic parameter and the non-dimensional parameters due to the initial stress in the equation of dispersion shows the significant effect on the propagation of love waves. The phase velocity of Love wave has been computed numerically and presented graphically. It is observed that the anisotropic parameter in the porous layer and the porosity of the layer both have the increasing effect but the initial stress field has an decreasing effect on the phase velocity of Love wave.

FORMULATION OF THE PROBLEM AND ITS FINITE DIFFERENCE APPROXIMATION

Formulation

We consider a model consisting of fluid saturated anisotropic poroelastic layer of finite thickness under compressive initial stress; the x-axis is chosen parallel to the layer in the direction of propagation of surface wave and the z-axis is taken vertically downward.

Neglecting the viscosity of the fluid and the body force, Biot's dynamical equations for the fluid-saturated anisotropic porous layer under compressive initial stress are given by [1,2]

$$\frac{\partial s_{11}}{\partial x}+\frac{\partial s_{12}}{\partial y}+\frac{\partial s_{13}}{\partial z}-P\left(\frac{\partial w_3}{\partial y}+\frac{\partial w_2}{\partial z}\right)=\frac{\partial^2}{\partial t^2}\left(\rho_{11}u_1+\rho_{12}U_1\right)$$

$$\frac{\partial s_{21}}{\partial x}+\frac{\partial s_{22}}{\partial y}+\frac{\partial s_{23}}{\partial z}-P\frac{\partial w_3}{\partial x}=\frac{\partial^2}{\partial t^2}\left(\rho_{11}u_2+\rho_{12}U_2\right)$$

$$\frac{\partial s_{31}}{\partial x}+\frac{\partial s_{32}}{\partial y}+\frac{\partial s_{33}}{\partial z}+P\frac{\partial w_2}{\partial z}=\frac{\partial^2}{\partial t^2}\left(\rho_{11}u_3+\rho_{12}U_3\right) \quad (1)$$

$$\left.\begin{aligned}\frac{\partial s}{\partial x} &= \frac{\partial^2}{\partial t^2}\left(\rho_{21}u_1 + \rho_{22}U_1\right),\\ \frac{\partial s}{\partial y} &= \frac{\partial^2}{\partial t^2}\left(\rho_{12}u_2 + \rho_{22}U_2\right),\\ \frac{\partial s}{\partial z} &= \frac{\partial^2}{\partial t^2}\left(\rho_{12}u_3 + \rho_{22}U_3\right)\end{aligned}\right\} \tag{2}$$

where $P = -S_{11}$, the initial stress along x axis, $S_{ij},(i,j=1,2,3)$, are the components of stress tensor in the solid skeleton, $S(=-fp)$ is the reduced pressure of the fluid (p is the pressure in the fluid, and f is the porosity of the porous layer), u_i are the components of the displacement vector of the solid and U_i are those of fluid of the porous aggregate, $w_i,(i=2,3)$ are the angular displacement vectors. The dynamic coefficients, $\rho_{11}, \rho_{12}, \rho_{22}$ take into account of the inertia effects of the moving fluid and are related to the densities of the solid ρ_s the fluid ρ_f and the layer ρ' by the equation

$$\rho_{11} + \rho_{12} = (1-f)\rho_S, \quad \rho_{12} + \rho_{22} = f\rho_f,$$

so that the mass density of the aggregate is

$$\rho' = \rho_{11} + 2\rho_{12} + \rho_{22} = \rho_s + f\left(\rho_f - \rho_S\right).$$

Also the dynamic coefficients, moreover, obey the inequalities $\rho_{11} > 0, \rho_{12} \le 0, \rho_{22} > 0, \rho_{11}\rho_{22} - \rho_{12}^2 > 0.$

The stress-strain relations for the fluid-saturated anisotropic porous layer under initial stress are given by

$$\left.\begin{aligned}
s_{11} &= (A+P)e_{11} + (A+P-2N)(e_{22}+e_{33}) + Q\varepsilon \\
s_{22} &= (A+P-2N)e_{11} + (A+P)e_{22} \\
&\quad + (A+P-2N)e_{33} + Q\varepsilon \\
s_{11} &= (A+P-2N)(e_{11}+e_{22}) + (A+P)e_{33} + Q\varepsilon \\
s_{12} &= 2Ne_{12}, s_{13} = 2Le_{13}, s_{23} = 2Le_{23}, \\
s &= Qe + R\varepsilon
\end{aligned}\right\} \quad (3)$$

where $e_{ij} = \frac{1}{2}(u_{i,j} + u_{j,i})$ and $e = \nabla \cdot u$ and $\varepsilon = \nabla \cdot U$

are the corresponding dilatations which opposite in sign, A, N, L correspond to the familiar Lame's constants, R is the amount of pressure required on the fluid to force a certain volume of the fluid into the aggregate while the total volume remains constants and Q is the coefficient of coupling between the volume change of the solid and that of the fluid.

Also the angular displacements are given by

$$w_2 = \frac{1}{2}\left(\frac{\partial u_1}{\partial z} - \frac{\partial u_3}{\partial x}\right) \text{ and } w_3 = \frac{1}{2}\left(\frac{\partial u_2}{\partial x} - \frac{\partial u_1}{\partial y}\right)$$

For the propagation of Love waves along x axis, using conventional conditions

$$i.e. \;\; \boldsymbol{u} = (0, u_2, 0) \text{ and } \boldsymbol{U} = (0, U_2, 0)$$

where $u_2 = v(x, z, t)$ and $U_2 = V(x, z, t)$ the equations of motion given by (1) and (2) are reduced to the form

$$N'\left(\frac{\partial^2 v}{\partial x^2}\right)+L\left(\frac{\partial^2 v}{\partial z^2}\right)=\frac{\partial^2}{\partial t^2}[\rho_{11}v+\rho_{12}V] \tag{4}$$

$$\frac{\partial^2}{\partial t^2}[\rho_{12}v+\rho_{22}V]=0 \tag{5}$$

Where $N' = N(1-\xi)$ and $\xi = P/2N$ is the non-dimensional parameters due to the initial stress.

Eliminating the component of liquid displacement V from (4) and (5) we have

$$N'\left(\frac{\partial^2 v}{\partial x^2}\right)+L\left(\frac{\partial^2 v}{\partial z^2}\right)=d'\frac{\partial^2 v}{\partial t^2} \tag{6}$$

where

$$d' = \rho_{11} - \frac{\rho_{12}^2}{\rho_{22}} \tag{7}$$

From the Equation (6) it can be seen that the velocities of shear waves in the porous medium in x and z directions are $\sqrt{N'/d'}$ and $\sqrt{L/d'}$ respectively and from (7), d′ is less than ρ_{11} and hence, in turn, less than ρ', the density of the elastic layer for all values of $\rho_{12} \neq 0$. This shows that the velocities of shear waves in x and z directions in the fluid saturated porous layer will be more those of corresponding elastic layer.

Finite Difference Approximation

Equation (6) can be written as

$$\gamma\left(\frac{\partial^2 v}{\partial x^2}\right)+\left(\frac{\partial^2 v}{\partial z^2}\right)=\frac{\gamma}{\alpha^2}\frac{\partial^2 v}{\partial t^2} \tag{8}$$

where $\gamma = \varsigma(1-\xi), \varsigma = N/L$ and $\alpha = \sqrt{N'/d'}$, the velocity of shear wave in the porous medium in x direction.

The shear wave velocity in the x-direction may be expressed as

$$\alpha = \beta_a\sqrt{\frac{(1-\xi)}{d}} \tag{9}$$

where $d = \gamma_{11} - \gamma_{12}^2/\gamma_{22}$ and $\beta_a = \sqrt{N/\rho'}$, the velocity of the shear wave in the corresponding non-porous, initial stress free anisotropic elastic medium along the direction of x. Also,

$$\gamma_{11} = \rho_{11}/\rho',\ \gamma_{12} = \rho_{12}/\rho,\ \gamma_{22} = \rho_{22}/\rho' \tag{10}$$

are the non-dimensional parameters for the material of the porous layer [1,2].

To improve the accuracy we have considered here the higher order finite difference scheme for spatial derivatives as

$$\frac{\partial^2 v}{\partial x^2}\approx\frac{1}{h^2}\left[a_0 v_{0,0}^0+\sum_{m=1}^{M}a_m\left(v_{m,0}^0+v_{-m,0}^0\right)\right] \tag{11}$$

$$\frac{\partial^2 v}{\partial z^2} \approx \frac{1}{h^2}\left[a_0 v_{0,0}^0 + \sum_{m=1}^{M} a_m\left(v_{0,m}^0 + v_{0,-m}^0\right)\right] \tag{12}$$

As generally higher order finite difference on temporal derivatives scheme requires large space in the computer memory and usually unstable, 2nd order finite difference scheme is used for temporal derivatives as

$$\frac{\partial^2 v}{\partial t^2} \approx \frac{1}{\tau^2}\left[-2v_{0,0}^0 + \left(v_{0,0}^1 + v_{0,0}^{-1}\right)\right] \tag{13}$$

where $v_{m,j}^n = v(x+mh, z+jh, t+n\tau)$, h is the grid size and τ is the time step.

Using (11)-(13) into (8) we have

$$a_0(1+\gamma)v_{0,0}^0 + \sum_{m=1}^{M} a_m\left[\gamma\left(v_{m,0}^0 + v_{-m,0}^0\right) + \left(v_{0,m}^0 + v_{0,-m}^0\right)\right]$$
$$= \frac{\gamma h^2}{\tau^2 \alpha^2}\left[-2v_{0,0}^0 + \left(v_{0,0}^1 + v_{0,0}^{-1}\right)\right] \tag{14}$$

Using the plane wave theory, let us consider

$$v_{m,j}^n = \mathrm{e}^{\mathrm{i}\left[k_x(x+mh)+k_z(z+jh)-\omega(t+n\tau)\right]} \tag{15}$$

Substituting (15) into (14) and simplifying, we have

$$\frac{1}{2}a_0(1+\gamma)+\sum_{m=1}^{M}a_m\left[\gamma\cos(mkh\cos\theta)+\cos(mkh\sin\theta)\right]$$
$$=\gamma r^{-2}\left[-1+\cos(\omega\tau)\right] \tag{16}$$

Where $r=\frac{\tau\alpha}{h}$, $k_x=k\cos\theta$, $k_z=k\sin\theta$, θ, being the propagation direction angle of the plane wave.

Using the Taylor series expansion for cosine functions, we have from (16)

$$\sum_{m=1}^{M}a_m\left\{(1+\gamma)+\sum_{j=1}^{\infty}(-1)^j\frac{\left[\gamma\cos^{2j}\theta+\sin^{2j}\theta\right](mkh)^{2j}}{(2j)!}\right\}$$
$$+\frac{1}{2}a_0(1+\gamma)=\gamma\sum_{j=1}^{\infty}(-1)^j\frac{r^{2j-2}(kh)^{2j}}{(2j)!} \tag{17}$$

Comparing the coefficients of k^{2j}, we get,

$$a_0+2\sum_{m=1}^{M}a_m=0 \tag{18}$$

$$\sum_{m=1}^{M}a_m\left(\gamma\cos^{2j}\theta+\sin^{2j}\theta\right)(m)^{2j}=\gamma r^{2j-2}, \tag{19}$$

where $j=1,2,\cdots,M$.

This equation indicates that the coefficients a_m are the function of θ. To obtain a single set of coefficients, an optimal angle has to be chosen.

We solve the equation (19) to get a_m by using $\theta = \pi / 4$ and then a_0 can be obtained from (18).

ERRORS AND ACCURACY

The error function of Equation (17) can be written as

$$\varepsilon = \left| \sum_{j=M+1}^{\infty} \frac{2(-1)^j}{(2j)!} \left\{ \sum_{m=1}^{M} a_m \left[\gamma \cos^{2j} \theta + \sin^{2j} \theta \right] (m)^{2j} - \gamma r^{2j-2} \right\} k^{2j} h^{2j-2} \right| \tag{20}$$

Since the minimum power of h in the error function (20) is 2M, the accuracy of this finite difference scheme is 2M when we use 2nd order time domain finite difference scheme and 2M-th order space domain finite difference scheme. The increase in M may decrease the magnitude in errors but may not increase the order of accuracy. As we have used the finite difference scheme in both time domain and space domain, the wave Equation (8) can be solved in both time domain and space domain simultaneously by (14).

DISPERSION ANALYSIS

Let us define a parameter δ to describe the dispersion of Finite difference by using Equation (17) as follows:

$$\delta = \frac{v_{FD}}{\beta_a} = \frac{1}{r\pi h_x} \cdot \sqrt{\frac{1-\xi}{d}} \sin^{-1}\left(\sqrt{\frac{r^2}{\varsigma(1-\xi)}} X \right) \tag{21}$$

where

$$X = \sum_{m=1}^{M} a_m \left[\varsigma (1-\xi) \sin^2 (m\pi h_x \cos\theta) + \sin^2 (m\pi h_x \sin\theta) \right] \tag{22}$$

If δ is equal to 1, then there is no dispersion. However, if δ is far from 1, a large dispersion will occur. In calculating δ, $d \to 0$, the grid points per wavelength, ranges from 0.04 to 0.5 and the variation of θ is from 0 to $\pi/4$.

The presence of d, the porosity parameter, $\varsigma = N/L$, the non-dimensional anisotropic parameter and $\xi = P/2N$, the non-dimensional parameters due to the initial stress shows that the dispersion curve are affected by these also.

It may be noted that (1-d) gives the fraction of porosity in the layer. If the layer is non-porous then $f \to 0$ and hence $\rho_s \to \rho'$ and we find $\gamma_{11} + \gamma_{12} \to 1$

and $\gamma_{12} + \gamma_{22} \to 0$ which leads to $\gamma_{11} - \frac{\gamma_{12}^2}{\gamma_{22}} \to 1$ i.e. $d \to 1$.

Again if $f \to 1$ then $\rho_f \to \rho'$ and the layer becomes a fluid, and in that case the shear wave velocity in the layer cannot exist which so happens when $d \to 0$.

Thus we have the following:

I. $d \to 1$, when the layer is non-porous solid;
II. $d \to 0$, when the layer tends to be fluid;
III. $0 < d < 1$, when the layer is porous.

Here we study the dispersion curves for different grid points per wave length, velocities, porosity, anisotropic parameters, non-dimensional parameters due to the initial stress and time steps.

NUMERICAL CALCULATION AND DISCUSSIONS

The numerical calculation of the Equation (21) has been done for different values of the parameters g, d and by taking, $\alpha = 1500$. The phase velocity v_{FD} / β_a of Love wave from the Equation (21) versus h_x, the grid points per wavelength, has been computed for different values of

$$d = 1, 0.95, 0.90, 0.85, 0.80, \cdots, 0.45,$$

$$\xi = P/2N = 0, 0.2, 0.4, 0.6, 0.8, \quad \varsigma = N/L = 1, 2,$$

the propagation angle $\theta = 0, \pi/16, \pi/8, 3\pi/16, \pi/4$ and $t = 0.0005, 0.001, 0.0015, 0.002, 0.0025, 0.003$.

Figures 1-3 display the dispersion curves of Love waves with respect to different grid points per wave length, at different values of M in a homogeneous non porous initial stress free elastic solid, porous isotropic and anisotropic layer under the influence of initial stress respectively. It is found that dispersion is more for the lower values of M and less for higher values of M. Here also it is observed that the increase in porosity and anisotropy leads to the increase in the magnitude of the phase velocity of Love waves whereas the increase in initial stress parameter leads to decrease in the phase velocity of Love waves.

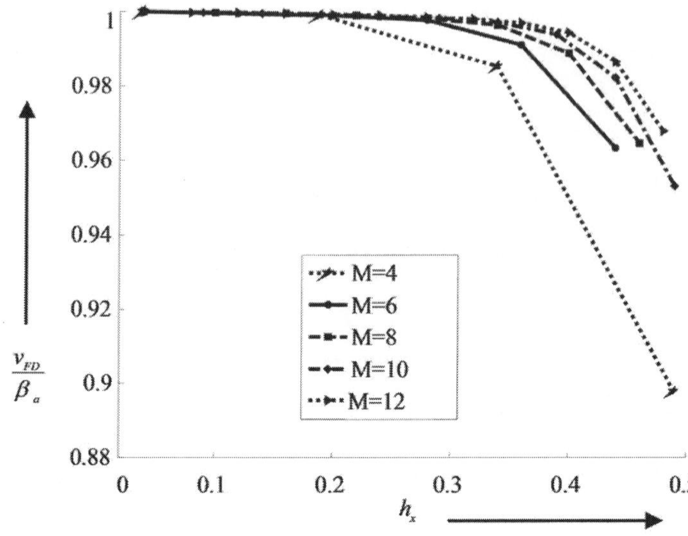

Figure 1: Dispersion curves for love waves for different values of M when $\varsigma = 1$, d=1 and $\xi = 0.0$.

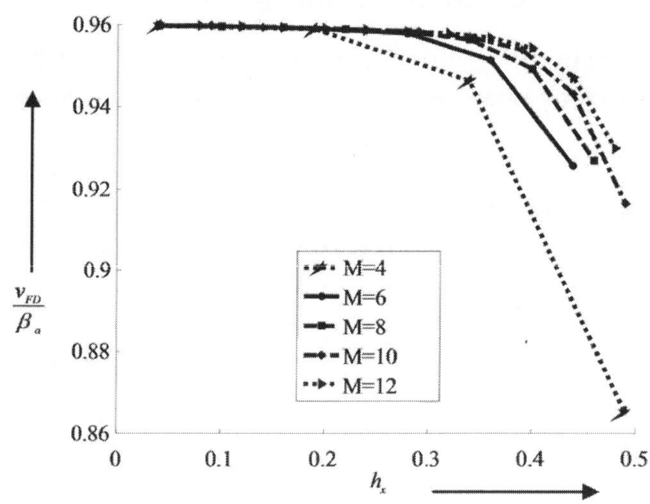

Figure 2: Dispersion curves for love waves for different values of M when $\varsigma = 1$, d=0.8 and $\xi = 0.2$.

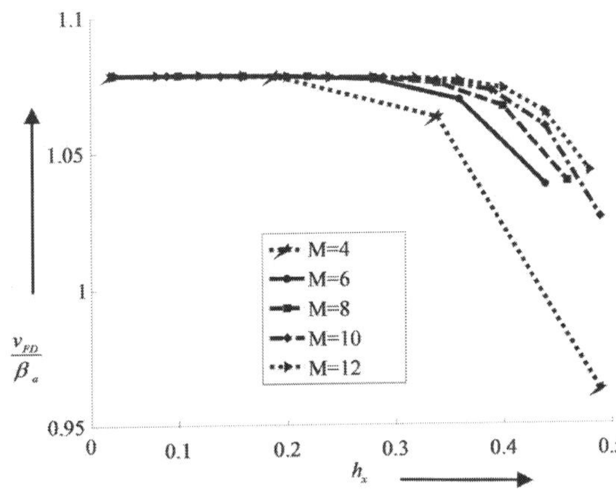

Figure 3: Dispersion curves for love waves for different values of M when $\varsigma = 2$, d=0.8 and $\xi = 0.2$.

Figure 4 shows the effect of porosity in the dispersion of phase velocity of Love waves with respect to different grid points per wave length in a homogeneous, isotropic porous layer without initial stress field and under the influence of initial stress. It is observed that as the porosity increases, that is, as the value of d decreases, the phase velocity of Love wave increases and decreases as the initial stress field increases.

Figures 5 and 6 show the effect of initial stress field in the propagation of Love waves with respect to different grid points per wave length in a isotropic and an-isotropic non-porous elastic solid and fluid saturated porous layer. As the anisotropy increases, the phase velocity of Love wave increases and the phase velocity of Love wave decreases as the initial stress field increases.

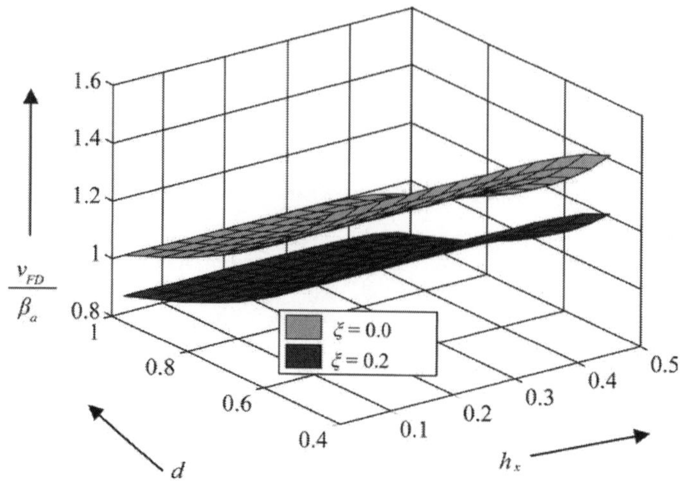

Figure 4: Dispersion curves for love waves for different values of porosity d and initial stress ξ when ς =1.

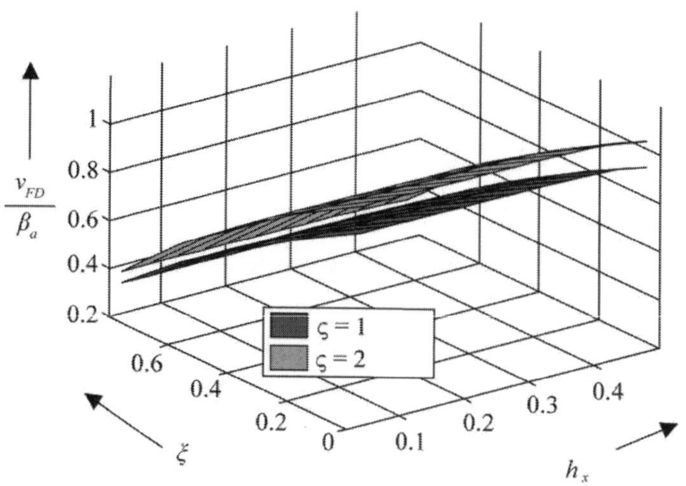

Figure 5: Dispersion curves for love waves for different values of initial stress ξ and anisotropic parameter ς when d = 1.

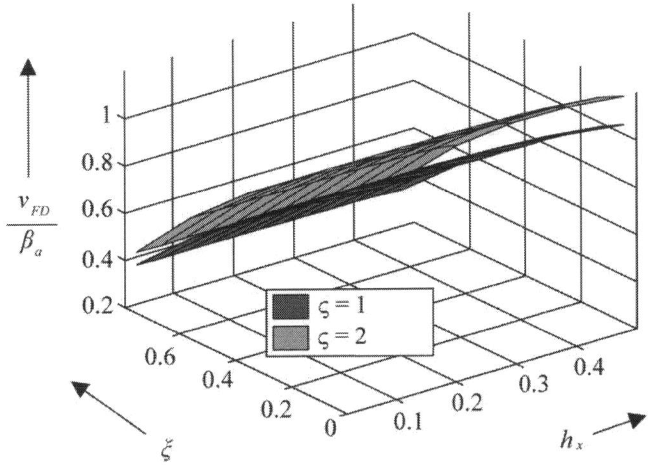

Figure 6: Dispersion curves for love waves for different values of initial stress ξ and anisotropic parameter ς when d=0.8.

Figures 7 and 8 display the dispersion curves of Love waves with respect to different grid points per wave length at different angles of propagation in a homogeneous fluid saturated porous layer and non-porous elastic solid layer for different values of initial stress respectively. It is observed that dispersion is more for the lower degree of propagation angles as compare to the higher degree of angle. Phase velocity is more in fluid saturated porous layer than in non-porous elastic solid layer. Later is the case discussed by Yang Liu et al. [20].

Figures 9 and 10 displays the dispersion curves of Love waves with respect to different grid points per wave length at different time steps in a homogeneous non porous elastic solid and non-homogeneous, anisotropic fluid saturated porous layer without initial stress field and under the influence of initial stress respectively. It is found that initially the dispersion is less and as time increases, the dispersion increases.

CONCLUSIONS

It is observed that the higher order time dependent finite difference method plays an important role in the propagation of Love waves in a porous layer under the influence of initial stress. Graphically we have shown that the dispersion curves of Love waves are less dispersed for higher order finite difference method. It has also been shown that initially the dispersion is less and as time increases, the dispersion increases. The significant effect of porosity, anisotropy and initial stress simultaneously in the propagation of Love waves in a porous layer has also been discussed. The phase velocity of Love wave is more in fluid saturated porous layer than in non-porous elastic solid layer. The anisotropy has an increasing effect whereas the initial stress field has a decreasing effect on the phase velocity of Love waves.

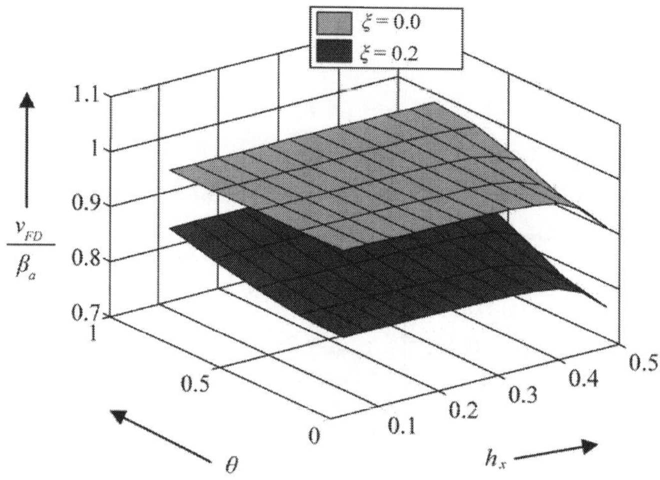

Figure 7: Dispersion curves for love waves at different propagation angle 0 for different values of initial stress ξ when anisotropic parameter $\varsigma = 1$ when d=1.

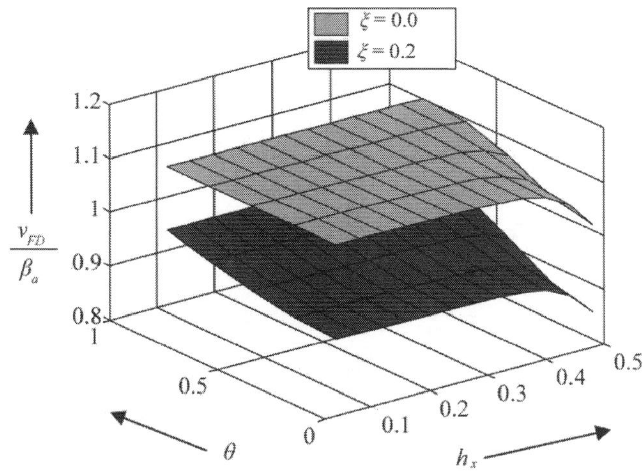

Figure 8: Dispersion curves for love waves at different propagation angle 0 for different values of initial stress ξ when anisotropic parameter $\varsigma = 1$ when d=0.8.

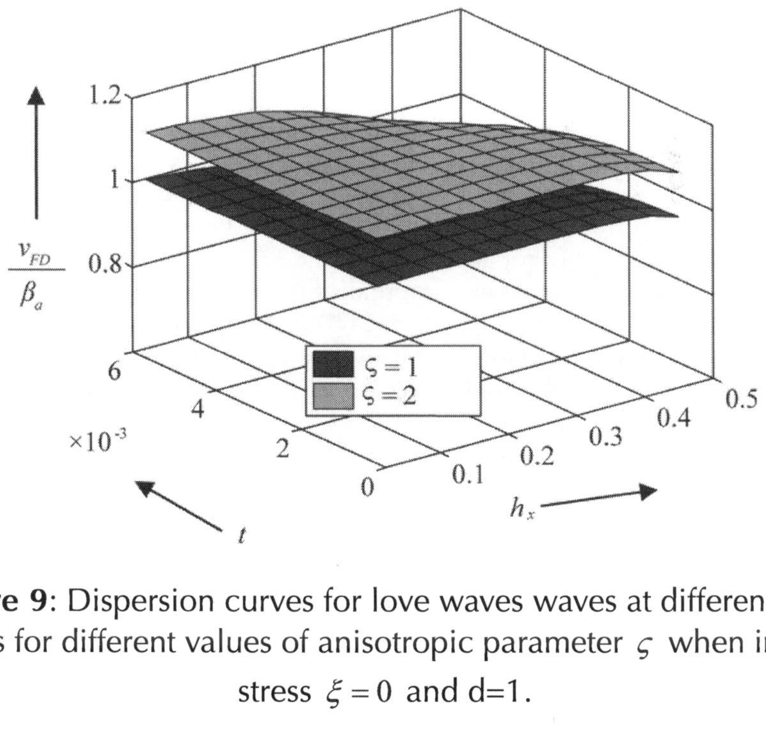

Figure 9: Dispersion curves for love waves waves at different time steps for different values of anisotropic parameter ς when initial stress $\xi = 0$ and d=1.

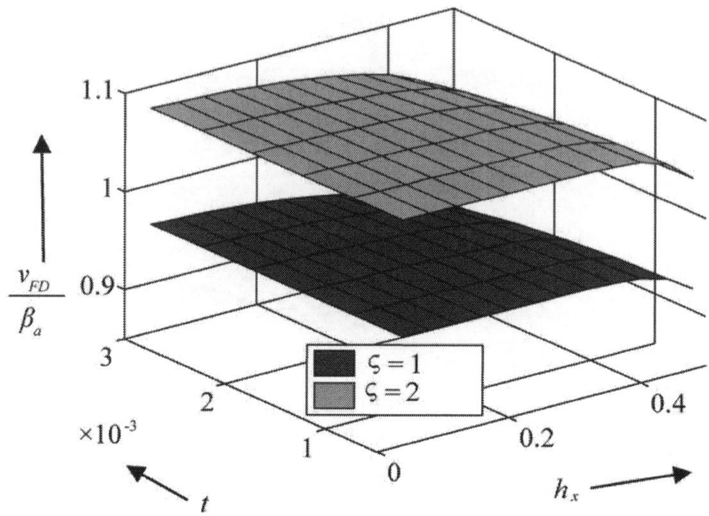

Figure 10: Dispersion curves for love waves waves at different time steps for different values of anisotropic parameter ς when initial stress $\xi = 0.2$ and d = 0.8.

REFERENCES

1. M. A. Biot, "Theory of Propagation of Elastic Solid in a Fluid-Saturated Porous Solid 1. Low-Frequence Range," Journal of the Acoustical Society of America, Vol. 28, 1956, pp. 168-178. doi:10.1121/1.1908239
2. M. A. Biot, "Theory of Propagation of Elastic Solid in a Fluid-Saturated Porous Solid II. Low-Frequence Range," Journal of the Acoustical Society of America, Vol. 28, 1956, pp. 179-191. doi:10.1121/1.1908241
3. M. A. Biot, "Surface Instability in Finite Anisotropic Elasticity under Initial Stress," Proceedings of the Royal Society A, Vol. 273, No. 1354, 1963, pp. 329-339. doi:10.1098/rspa.1963.0092
4. A. M. Abd-Alla, S. R. Mahmoud and M. I. R. Helmi, "Effect of Initial Stress and Magnetic Field on Propagation of Shear Wave in Non-Homogeneous Anisotropic Medium under Gravity Field," The Open Applied Mathematics Journal, Vol. 3, No. 1, 2009, pp. 58-65.doi:10.2174/1874114200903010058
5. S. Gupta, A. Chattopadhyay and D. K. Majhi, "Effect of Initial Stress on Propagation of Love Waves in an Anisotropic Porous Layer," Journal of Solid Mechanics, Vol. 2, No.1, 2010, pp. 50-62.

6. S. Gupta, A. Chattopadhyay, S. K. Vishwakarma and D. K. Majhi, "Influence of Rigid Boundary and Initial Stress on the Propagation of Love Wave," Applied Mathematics, Vol. 2, No. 5, 2011, pp. 586-594. doi:10.4236/am.2011.25078
7. J. Virieux, "SH-Wave Propagation in Heterogeneous Media: Velocity Stress Finite Difference Method," Geophysics, Vol. 49, No. 11, 1984, pp. 1933-1957. doi:10.1190/1.1441605
8. J. Virieux, "P-SV Wave Propagation in Heterogeneous Media: Velocity Stress Finite Difference Method," Geophysics, Vol. 51, No. 4, 1986, pp. 889-901. doi:10.1190/1.1442147
9. A Levander, "Fourth-Order Finite Difference P-SV Seismograms," Geophysics, Vol. 53, No. 11, 1998, pp. 1425- 1436. doi:10.1190/1.1442422
10. K. Hayashi and D. R. Burns, "Variable Grid Finite-Difference Modeling Including Surface Topography," 69th Annual International Meeting, SEG, Expanded Abstracts, Vol. 18, 1999, pp. 523-527.
11. R. W. Graves, "Simulating of Seismic Wave Propagation in 3D Elastic Media Using Staggered-Grid Finite Differences," Bulletin of the Seismological Society of America, Vol. 86, No. 4, 1996, pp. 1091-1106.
12. J. Kristek and P. Moczo, "Seismic Wave Propagation in Visco-Elastic Media with Material Discontinuities: A 3D Forth-Order Staggered-Grid Finite Difference Modeling," Bulletin of the Seismological Society of America, Vol. 93, No. 5, 2003, pp. 2273-2280.doi:10.1785/0120030023
13. E. H. Saenger and T. Bohlen, "Finite Difference Modeling of Viscoelastic and Anisotropic Wave Propagation Using the Rotated Staggered Grid," Geophysics, Vol. 69, No. 2, 2004, pp. 583-591. doi:10.1190/1.1707078
14. T. Bohlen and E. H. Saenger, "Accuracy of Heterogeneous Staggered-Grid Finite-Difference Modeling of Rayleigh Waves," Geophysics, Vol. 71, No. 4, 2006, pp. T109- T115. doi:10.1190/1.2213051
15. J. Kristek and P. Moczo, "On the Accuracy of the Finite-Difference Schemes: The 1D Elastic Problem," Bulletin of the Seismological Society of America, Vol. 96, No. 6, 2006, pp. 2398-2414. doi:10.1785/0120060031
16. E. Tessmer, "Seismic Finite Difference Modeling with Spatially Variable Time Steps," Geophysics, Vol. 65, No. 4, 2000, pp. 1290-1293. doi:10.1190/1.1444820
17. B. Finkelstein and R. Kastner, "Finite Difference Time Domain Dispersion Reduction Schemes," Journal of Computational Physics, Vol. 221, No. 1, 2007, pp. 422-438.doi:10.1016/j.jcp.2006.06.016
18. Y. Liu and M. K. Sen, "Advanced Finite-Difference Method for Seismic Modeling," Geohorizons, Vol. 14, No. 2, 2009, pp. 5-16.
19. Y. Liu and M. K. Sen, "A Practical Implicit Finite Difference Method: Examples from Seismic Modeling," Journal of Geophysics and Engineering, Vol. 6, No. 3, 2009, pp. 231-249. doi:10.1088/1742-2132/6/3/003
20. Y. Liu and M. K. Sen, "A New Time-Space Domain High-Order Finite Difference Method for the Acoustic Wave Equation," Journal of Computational Physics, Vol. 228, No. 23, 2009, pp. 8779-8806. doi:10.1016/j.jcp.2009.08.027

21. M. A. Dublain, "The Application of High-Order Differencing to the Scalar Wave Equation," Geophysics, Vol. 51, No. 1, 1986, pp. 54-66. doi:10.1190/1.1442040
22. Y. Liu and M. K. Sen, "Acoustic VTI Modeling with a Time-Space Domain Dispersion-Relation-Based FiniteDifference Scheme," Geophysics, Vol. 75, No. 3, 2010, pp. A11-A17.doi:10.1190/1.3374477
23. Y. Liu and W. Xiucheng, "Finite Difference Numerical Modeling with Even Order Accuracy in Two Phase Anisotropic Media," Applied Geophysics, Vol. 5, No. 2, 2008, pp. 107-114.doi:10.1007/s11770-008-0014-6
24. X. Zhu and G. A. McMechan, "Finite difference Modeling of the Seismic Response of fluid Saturated, Porous, Elastic Solid Using Biot Theory," Geophysics, Vol. 56, No. 4, 1991, pp. 424-435.
25. Y. Liu and M. K. Sen, "Scalar Wave Equation Modeling with Time-Space Domain Dispersion-Relation-Based Staggered-Grid Finite-Difference Schemes," Bulletin of the Seismological Society of America, Vol. 101, No. 1, 2011, pp. 141-159. doi:10.1785/0120100041.

CITATION

A. Ghorai and R. Tiwary, "Modeling of Surface Waves in a Fluid Saturated Poro-Elastic Medium under Initial Stress Using Time-Space Domain Higher Order Finite Difference Method," Applied Mathematics, Vol. 4 No. 3, 2013, pp. 469-476. doi: 10.4236/am.2013.43070.

Some Properties on the Error-Sum Function of Alternating Sylvester Series

Huiping Jing, Luming Shen
Science College of Hunan Agricultural University, Changsha, China

12

ABSTRACT

The error-sum function of alternating Sylvester series is introduced. Some elementary properties of this function are studied. Also, the hausdorff dimension of the graph of such function is determined.

INTRODUCTION

For any $x \in (0,1)]$, let $d_1 : d_1(x) \in N$ and $cT := T(x) \in (0,1)]$ be defined as

$$d_1(x) = \left[\frac{1}{x}\right], (x) := \frac{1}{d_1(x)} - x, T(0) := 0. \tag{1}$$

where $[\ ]$ denote the integer part. And we define the sequence $\{d_n(x), n \geq 2\}$ as follows:

$$d_n(x) = d_1\left(T^{n-1}(x)\right), \tag{2}$$

where T^n denotes the nth iterate of $T\left(T^0 = Id_{(0,1]}\right)$.

It is well known that from the algorithm (1), all $x \in (0,1]$ can be devel-opped uniquely into an infinite or finite series

$$x = \sum_{i\geq 1}(-1)^{i-1}\frac{1}{d_i(x)},$$
where $_{i+1}(x) \geq d_i(x)(d_i(x)+1)$. (3)

In the literature [2], (3) is called the Alternating Balkema-Oppenheim expansion of x and denoted by $x = [d_1(x),\cdots,d_n(x),\cdots]$ for short. From the algorithm, one can see that T maps irrational element into irrational element, and the series is infinite. While for rational numbers, in fact, we have $x \in (0,1)]$ is rational if and only if its sequence of digits $d_1(x),\cdots,$ is terminate or periodic, see [1-3].

For any $x \in (0,1)]$ and $n \geq 1$, define $\frac{p_n(x)}{q_n(x)} = \sum_{i=1}^{n}(-1)^{i-1}\frac{1}{d_i(x)}$. From the algorithm of (1), it is clear that

$$x = \frac{p_n(x)}{q_n(x)} + (-1)^n T^n(x). \tag{4}$$

For any $x \in (0,1)]$, let $x = [d_1(x),\cdots,d_n(x),\cdots]$ be its Alternating Sylvester expansion, then we have

$\left(d_{j+1}(x) \geq d_j(x)(d_j(x)+1\right)$ for any $j \geq 1$. On the other hand, any $\{d_j, j \geq 1\}$ of integer sequence satisfying

$\left(d_{j+1}(x) \geq d_j(x)(d_j(x)+1\right)$ for all $j \geq 1$ is a Sylvester admissible sequence, that is, there exists a unique $x \in (0,1)]$ such that $d_j(x) = d_j$ for all $j \geq 1$, see [9].

The behaviors of the sequence $d_n(x)$ are of interest and the metric and ergodic properties of the sequence $\{d_n(x), n \geq 1\}$ and T have been investigated by a number of authors, see [1-3].

For any $x \in (0,1)]$, define

$$S(x) := \sum_{n=1}^{+\infty}\left(x - \frac{p_n(x)}{q_n(x)}\right), \tag{5}$$

and we call S(x) the error-sum function of Alternating Sylvester series. By (4), since $d_{n+1}(x) \geq d_n(x)(d_n(x)+1)$ for all $n \geq 1$, then $|S(x)| \leq 1$ and S(x) is well defined. In this paper, we shall discuss some basic nature of S(x), also the Hausdorff dimension of the graph of S(x) is determined.

SOME BASIC PROPERTIES OF S(x)

In what follows, we shall often make use of the symbolic space.

For any $n \geq 1$, let $D_n = \{(\sigma 1, \sigma 2, \cdots \sigma_n) \in N^n : \sigma_{k+1} \geq \sigma_k(\sigma_k + 1)$ For all $1 \leq k \leq n\}$ Define

$$D = \bigcup_{n=0}^{\infty} D_n, (D_0 := \varnothing).$$

For any $\sigma = (\sigma 1, \sigma 2, \cdots \sigma_n) \in D_n$, write

$$A_\sigma = \frac{1}{\sigma_1} - \frac{1}{\sigma_2} + \cdots + (-1)^{n-1}\frac{1}{\sigma_n}, \tag{6}$$

$$B_\sigma = \frac{1}{\sigma_1} - \frac{1}{\sigma_2} + \cdots + (-1)^{n-1}\frac{1}{\sigma_n + 1}. \tag{7}$$

We use J_σ to denote the following subset of (0,1],

$$J_\sigma = \{x \in (0,1] : d_1(x) = \sigma_1, \\ d_2(x) = \sigma_2, \cdots, d_n(x) = \sigma_n\}. \quad (8)$$

From theorem 4.14 of [8], we have $J_\sigma = (A_\sigma, B_\sigma)]$ when n is even, and $J_\sigma = (B_\sigma, A_\sigma)]$ when n is odd. Finally, define

$$I = \{A_\sigma, B_\sigma, \sigma \in D_n, n \geq 1.\} \quad (9)$$

Lemma 1: For any $n \geq 1$ and $x \in (0,1)]$

$$\lim_{x \to 0^+} S(x) = 0; \quad (10)$$

$$-\frac{17}{30} \leq S(x) \leq 0; \quad (11)$$

$$S(x) = \sum_{i=1}^{n} \left(x - \frac{p_i(x)}{q_i(x)} \right) + (-1)^n S(T^n(x)). \quad (12)$$

Proof.1: Since $d_{j+1}(x) \geq d_j(x)(d_j(x)+1)$ and $d_1(x) \geq 1$, so when $n \geq 3$, we can get

$d_{n+1} \geq d_n^{\,2} > \cdots > d_2^{\,2^{n-1}}$, accordingly $d_n > d_2^{\,2^{n-1}} \geq \left(d_1^{\,2}(d_1+1)^2\right)^{n-2}$ we write $a(x) = d_1(x)^2 (d_1(x)+1)^2$, so $d_n > a(x)^{n-2}$. Now $d_{n+1}(x) = \left[\frac{1}{T^n(x)}\right]$ implies

$$\frac{1}{d_{n+1}(x)+1} < T^n(x) \leq \frac{1}{d_{n+1}(x)}, \quad \text{for } 0 < T^n(x) \leq 1. \text{ Thus}$$

$$S(x)=\sum_{n=1}^{+\infty}(-1)^n T^n(x)\geq \sum_{n=1}^{+\infty}(-1)T^{2n-1}(x)$$

let $x\to 0^+$,

$$\geq \sum_{n=1}^{+\infty}(-1)\frac{1}{d_{2n}(x)}\geq -\frac{1}{d_2(x)}-\sum_{n=2}^{+\infty}\frac{1}{a(x)^{2n-2}}\geq -\frac{1}{\sqrt{a(x)}}-\frac{1}{a(x)^2-1},$$

we have $d_1(x)\to +\infty$ and $a(x)\to +\infty$, thus $S(x)\to 0$

2. From 1 we know that $d_{n+1}\geq d_n{}^2\cdots d_2{}^{2^{n-1}}$,

from the definition of $d_i(x)$ we also know that $d_1\geq 1$, so $d_2\geq d_1(d_1+1)\geq 2$,

$d_{n+1}>d_2{}^{2^{n-1}}\geq 4^{n-1}$, sthus

$$S(x)\geq \sum_{n=1}^{+\infty}(-1)\frac{1}{d_{2n}(x)}\geq -\frac{1}{2}-\sum_{n=2}^{+\infty}\frac{1}{4^{2n-2}}=-\frac{17}{30}$$

3. Since as $n>m$,

$$\frac{p_n(x)}{q_n(x)}-\frac{p_m(x)}{q_m(x)}=(-1)^m\frac{p_{n-m}\left(T^m(x)\right)}{q_{n-m}\left(T^m(x)\right)}.$$ Thus

$$S(x)=\sum_{i=1}^{\infty}\left(x-\frac{p_i(x)}{q_i(x)}\right)$$
$$=\sum_{i=1}^{n}\left(x-\frac{p_i(x)}{q_i(x)}\right)+\sum_{i=n+1}^{\infty}\left(x-\frac{p_n(x)}{q_n(x)}+\frac{p_n(x)}{q_n(x)}-\frac{p_i(x)}{q_i(x)}\right)$$
$$=\sum_{i=1}^{n}\left(x-\frac{p_i(x)}{q_i(x)}\right)+\sum_{i=n+1}^{\infty}\left[(-1)^n T^n(x)-(-1)^n\frac{p_{i-n}(T^n(x))}{q_{i-n}(T^n(x))}\right]$$
$$=\sum_{i=1}^{n}\left(x-\frac{p_i(x)}{q_i(x)}\right)+(-1)^n\sum_{i=1}^{\infty}\left[T^n(x)-\frac{p_i(T^n(x))}{q_i(T^n(x))}\right]$$
$$=\sum_{i=1}^{n}\left(x-\frac{p_i(x)}{q_i(x)}\right)+(-1)^n S(T^n(x)).$$

Let

$I' = I\setminus\{1\}$.

Proposition 2: For any $x\in I'$, if $x=[d_1(x),\cdots d_{2k+1}(x)]$, then $s(x)$ is left continuous but not right continuous. If $x=[d_1(x),\cdots d_{2k}(x)]$, then $s(x)$ is right continuous but not left continuous.

Proof: For any $n\geq 1$ and $\sigma\in D_n$, write $x_1 = A_\sigma$, $x_2 = B_\sigma$, where A_σ , B_σ are given by (6) and (7).

Case I, $n = 2k+1$, then

$$x_1=\frac{1}{\sigma_1}-\frac{1}{\sigma_2}+\cdots+\frac{1}{\sigma_{2k+1}} \tag{13}$$

$$x_2=\frac{1}{\sigma_1}-\frac{1}{\sigma_2}+\cdots+\frac{1}{\sigma_{2k+1}+1} \tag{14}$$

And $J_\sigma = (B_\sigma, A_\sigma]$. For any $x_1' \in J_\sigma$, since when $\sigma_{2k+1} = \sigma_{2k}(\sigma_{2k}+1)$,

$$\frac{1}{\sigma_1} - \frac{1}{\sigma_2} + \cdots - \frac{1}{\sigma_{2k}} + \frac{1}{\sigma_{2k+1}}$$
$$= \frac{1}{\sigma_1} - \frac{1}{\sigma_2} + \cdots - \frac{1}{\sigma_{2k}} + \frac{1}{\sigma_{2k}(\sigma_{2k}+1)}$$
$$= \frac{1}{\sigma_1} - \frac{1}{\sigma_2} + \cdots - \frac{1}{\sigma_{2k}+1}.$$

This situation is included in Case II, so we can take $\sigma_{2k+1} > \sigma_{2k}(\sigma_{2k}+1)$, and

$$x_1' = x1 - \frac{1}{\alpha} \text{for some}\, \alpha \geq (\sigma_{2k}+1) \quad \text{i.e. } x_1' = [\sigma_1, \cdots \sigma_{2k}, \sigma_{2k+1,} [\alpha] \cdots]$$

$$S(x_1') - S(x_1) = \sum_{i=1}^{2k+1} \left(x_1' - \frac{p_i(x_1')}{q_i(x_1')} \right) + \left(x_1' - \frac{p_{2k+2}(x_1')}{q_{2k+2}(x_1')} \right)$$
$$- \sum_{i=1}^{2k+1} \left(x_1 - \frac{p_i(x_1)}{q_i(x_1)} \right) + S(T^{2k+2}(x_1'))$$
$$= -\frac{2k+1}{\alpha} + T^{2k+2}(x_1') + S(T^{2k+2}(x_1'))$$

By (2),

$$\frac{1}{d_{n+1}(x)+1} < T^n(x) \leq \frac{1}{d_{n+1}(x)+1}, \text{for}\, 0 < T^n(x) \leq 1,$$
which implies

$$T^{n+1}(x) = \frac{1}{d_{n+1}(x)} - T^n(x) < \frac{1}{d_{n+1}(x)} - \frac{1}{d_{n+1}(x)+1} = \frac{1}{d_{n+1}(x)(d_{n+1}(x)+1)}$$

and

$$0 < T^{2n+2}(x_1') < \frac{1}{\alpha(\alpha+1)}.$$

Let $\alpha \to \infty$, we get $T^{2k+2}(x_1') \to 0$ and $S\left(T^{2k+2}(x_1')\right) \to 0$, thus

$$\lim_{x'_1 \to x^-_1} S(x_1') = S(x_1),$$

and this implies $S(x)$ is left continuous at x_1.

Let

$x_1'' = x_1 + \frac{1}{\alpha}$ for some

$\alpha \geq \left(\sigma_{2k+1,} - 1\right)\sigma_{2k+1,}\left((\sigma_{2k+1,} - 1)\sigma_{2k+1,} + 1\right),$

i.e. $e_1'' = \left[\sigma_1, \cdots, \sigma_{2k}, \sigma_{2k+1,} - 1\left(\sigma_{2k+1,} - 1\right)\sigma_{2k+1,}[\alpha], \cdots\right],$

then

$$\begin{aligned}
&S(x_1'') - S(x_1) \\
&= \sum_{i=1}^{2k}\left(x_1'' - \frac{p_i(x_1'')}{q_i(x_1'')}\right) + \left(x_1'' - \frac{p_{2k+1}(x_1'')}{q_{2k+1}(x_1'')}\right) \\
&\quad + \left(x_1'' - \frac{p_{2k+2}(x_1'')}{q_{2k+2}(x_1'')}\right) + \left(x_1'' - \frac{p_{2k+3}(x_1'')}{q_{2k+3}(x_1'')}\right) - S\left(T^{2k+3}(x_1'')\right) \\
&\quad - \left(\sum_{i=1}^{2k}\left(x_1 - \frac{p_i(x_1)}{q_i(x_1)}\right) + \left(x_1 - \frac{p_{2k+1}(x_1)}{q_{2k+1}(x_1)}\right)\right) \\
&= \frac{2k+2}{\alpha} - \frac{1}{(\sigma_{2k+1} - 1)\sigma_{2k+1}} - T^{2k+3}(x_1'') - S\left(T^{2k+3}(x_1'')\right).
\end{aligned}$$

Let $\alpha \to +\infty$, we have

$$\lim_{x'_2 \to x^+_2} S(x''_1) = S(x_1) - \frac{1}{\left(\sigma_{2k+1,} - 1\right)\sigma_{2k+1}}$$

and this implies $S(x)$ is not right continuous at x_1. For

$$x_2 = \frac{1}{\sigma_1} - \frac{1}{\sigma_2} + \cdots + \frac{1}{\sigma_{2k+1} + 1}, \tag{15}$$

following the same line as above, we have

$$\lim_{x'_2 \to x^+_2} S(x''_2) = S(x_2) - \frac{1}{\sigma_{2k+1}\left(\sigma_{2k+1,} - 1\right)}.$$

Case II $n = 2k$

Let

$$y_1 = \frac{1}{\sigma_1} - \frac{1}{\sigma_2} + \cdots - \frac{1}{\sigma_{2k}} \tag{16}$$

$$y_2 = \frac{1}{\sigma_1} - \frac{1}{\sigma_2} + \cdots - \frac{1}{\sigma_{2k} + 1} \tag{17}$$

Following the same line as above, we have

$$\lim_{y'_1 \to y_1^-} S(y'_1) = S(y_1) + \frac{1}{\left(\sigma_{2k+1,} - 1\right)\sigma_{2k}},$$

$$\lim_{y'_2 \to y_2^-} S(y'_2) = S(y_2) + \frac{1}{\sigma_{2k}\left(\sigma_{2k+1,} - 1\right)},$$

and $S(y_1), S(y_2)$ is right continuous.

Corollary 3. For any $n \geq 1$ and $\sigma \in D_n$, write $\alpha_1 = \max\{A_\sigma, B_\sigma\}$, $\alpha_2 = \min\{A_\sigma, B_\sigma\}$. Then for any $x \in J_\sigma$, if $n = 2k+1$, then

$$S^*(\alpha_2) < s(x) \leq s(\alpha_1),$$

Where $S^*(\alpha_2) = S(\alpha_2) - \dfrac{1}{\sigma_{2k+1}(\sigma_{2k+1,} + 1)}.$

From the corollary, for any $\sigma \in D_n$

$$\sup_{x,y \in J_\sigma} |s(x) - s(y)| = \frac{n}{\sigma_n(\sigma_n + 1)} = n\lambda(J_\sigma)$$

where $\lambda(J_\sigma)$ is the Lebesgue measure of J_σ.

Theorem 4. $s(x)$ is continuous on $(0,1] \setminus I'$.

Proof: For any $x \in (0,1] \setminus I'$ and $x \neq 1$,, let $x = (d_1(x), \cdots, d_n(x), \cdots)$ be its Alternating Sylvester expansion. For any $n \geq 1$, write $\sigma_n = (d_1(x), \cdots, d_n(x))$. By (Corollary 3), for any $y \in J_\sigma(n)$, we have

$$|s(x) - s(y)| \leq (n)\lambda(J_\sigma(n)) \to 0, \text{as} \to \infty.$$

Write $I_0 = \{c_\sigma\}$s, where $C_\sigma = \dfrac{1}{\sigma_1} - \dfrac{1}{\sigma_2} + \cdots + \dfrac{1}{\sigma_{2k+1}}$

Theorem 5: If $0 < a < b < 1, S(a) < y < s(b)$, then there exists $c \in (a,b) \setminus \{I_0\}$, such that $s(c) = y$.

Proof. Set $g(x) = s(x) = y$, then $g(x)$ has the same continuity as $s(x)$. Write

$$E = \{x | g(x) < 0, x \in [a,b]\}, \quad x_0 = \sup E.$$

trivially, $a \in E$, then the set is well defined.

If $b=[\sigma_1,\sigma_2,\cdots\sigma_{2k+1}]$, then by the left continuity of $S(b)$, we have

$$\lim_{x\to b^-} g(x)=g(b)>0,$$

As a result, there exists a $\delta>0$ such that for any $x\in(b-\delta_1,b), g(x)>0$.

If $b=[\sigma_1,\sigma_2,\cdots\sigma_{2k}]$, since $g(b)$ is not left continuous, then $\exists\delta_2>0$ such that for any $x\in(b-\delta_2,b)$, $g(x)>0$, that is $x_0\neq b$.

Following the same line as above, we can prove $x_0>a$.

Now we shall prove that $g(x_0)\leq 0$. We can choose $x_n\in E$ such that

$x_n\to x^-_0$, if $x_0=[\sigma_1,\sigma_2,\cdots\sigma_{2k+1}]$, then $g(x_0)=\lim\limits_{x_n\to x^-_0} g(x_n)\leq 0$,

If $x_0=[\sigma_1,\sigma_2,\cdots\sigma_{2k+1}]$, then

$$g(x_0)+\frac{1}{(\sigma_{2k}-1)\sigma_{2k}}=\lim_{x_n\to x^-_0} g(x_n)\leq 0,$$

In both case $g(x_n)\leq 0$. Following the same line as above, we can prove $g(x_0)=0$, and $x_0\neq[\sigma_1,\sigma_2,\cdots\sigma_{2k+1}]$.

Therefore, there exists $c\in(a,b)\setminus\{I_0\}$, such that $S(c)=y$.

Theorem 6: $\int_0^1 S(x)dx+\sum_{k=0}^{+\infty}\frac{1}{k(k+1)S(x)dx}=\frac{9-\pi^2}{6}$, and

$\int_0^1 S(x)dx=-0.1250.$

Proof

$$\int_0^1 S(x)\mathrm{d}x = \sum_{d_1=1}^{+\infty}\int_{\frac{1}{d_1+1}}^{\frac{1}{d_1}} S(x)\mathrm{d}x$$

$$= \sum_{d_1=1}^{+\infty}\int_{\frac{1}{d_1+1}}^{\frac{1}{d_1}} \left(\left(x-\frac{1}{d_1}\right)-S(T(x))\right)\mathrm{d}x$$

$$= \sum_{d_1=1}^{+\infty}\int_{\frac{1}{d_1+1}}^{\frac{1}{d_1}} x\mathrm{d}x - \sum_{d_1=1}^{+\infty}\int_{\frac{1}{d_1+1}}^{\frac{1}{d_1}} \frac{1}{d_1}\mathrm{d}x - \sum_{d_1=1}^{+\infty}\int_{\frac{1}{d_1+1}}^{\frac{1}{d_1}} S(T(x))\mathrm{d}x$$

Let $Tx = u = \frac{1}{d_1(x)} - x$, then $du = -dx$ thus

$$\int_0^1 S(x)\mathrm{d}x = \frac{1}{2}\sum_{d_1=1}^{+\infty}\left(\frac{1}{d_1^2}-\frac{1}{(d_1+1)^2}\right)$$

$$-\sum_{d_1=1}^{+\infty}\frac{1}{d_1^2}+\sum_{d_1=1}^{+\infty}\frac{1}{d_1(d_1+1)}-\sum_{d_1=1}^{+\infty}\int_0^{\frac{1}{d_1(d_1+1)}} S(u)\,du$$

thus,

$$\int_0^1 S(x)dx + \sum_{k=0}^{+\infty}\frac{1}{k(k+1)S(x)dx = \frac{3}{2}-\sum_{d_1=1}^{+\infty}\frac{1}{d_1^2} = \frac{9-\pi^2}{6}}.$$

Through the MATLAB program we can get the definite integration $\int_0^1 S(x)dx = -0.1250.$

HAUSDORFF DIMENSION OF GRAPH FOR $s(x)$

Write

$$Gr(S) = \{(x, S(x)), x \in (0,1]\}.$$

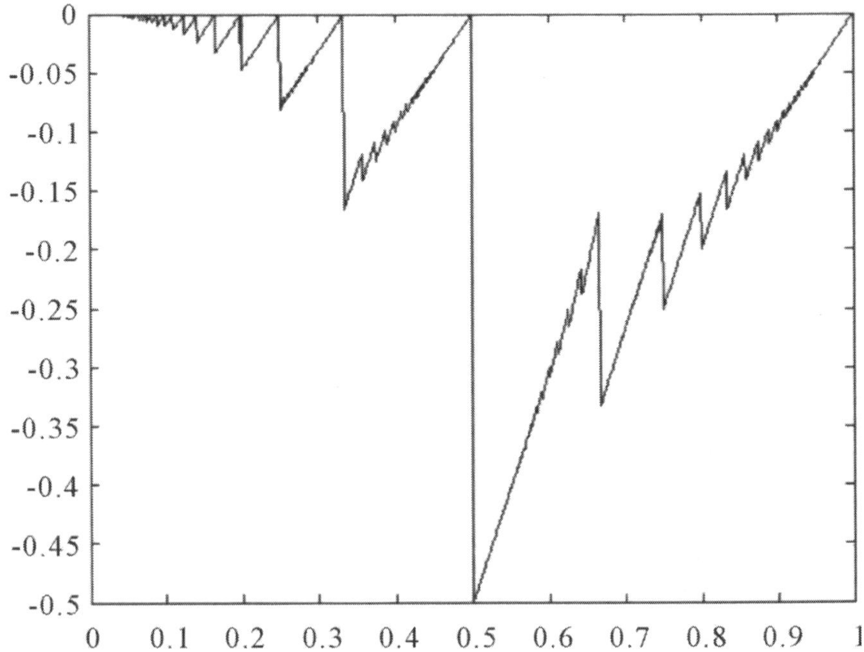

Figure 1: The graph of S(x).

Theorem 7: $\dim_H Gr(S) = 1$.

Proof: For any $n \geq 1$, $\{J_\sigma \times S(J_\sigma), \sigma \in D_n\}$ is a covering of $Gr(S)$. From (Cor 3), $J_\sigma \times S(J_\sigma)$ can be covered by n squares with side of length $\lambda(J_\sigma)$. For any $\varepsilon > 0$,

$$\begin{aligned} H^{1+\varepsilon}\left(Gr(S)\right) &\leq \liminf_{n\to\infty} \sum_{\sigma\in D_n} n\left(\sqrt{2}\right)^{1+\varepsilon} \left(\lambda(J_\sigma)\right)^{1+\varepsilon} \\ &\leq \liminf_{n\to\infty} n\left(\sqrt{2}\right)^{1+\varepsilon} 2^{-n\varepsilon} \sum_{\sigma\in D_n} n\left(\sqrt{2}\right)^{1+\varepsilon} \\ &= \liminf_{n\to\infty} n\left(\sqrt{2}\right)^{1+\varepsilon} 2^{-n\varepsilon} = 0. \end{aligned}$$

Thus, $\dim_H Gr(S) \leq 1$

Since

$$|proj(x,S(x))| - proj(y,S(y)) \leq d(x,S(x),(y,S(y)) \text{ then}$$

$$1 = \lambda((0,1]) = H^1(0,1] = H^1\left(proj\left(G_r(S)\right)\right) \leq^1 (G_r(S)),$$

So $\dim_H Gr(S) = 1$.

ACKNOWLEDGEMENTS

This work is supported by the Hunan Education Department Fund (11C671).

REFERENCES

1. S. Kalpazidou, A. Knopfmacher and J. Knopfmacher, "Lü roth-Type Alt entations for Real Numbers," Ac , No. 4, 1990, pp. No. 3, 1991, pp. 319-325. p. 311- ol. 8, No. 2, 1996, pp. 331-346. B44, 1992, pp. 17-28.
2. S. Kalpazidou, A. Knopfmacher and J. Knopfmacher, "Metric Properties of Alternating Lüroth Series," Potugaliae Mathematica, Vol. 48,
3. J. Barrionuevo, M. Burton-Robert, Dajani-Karma and C. Kraaikamp, "Ergodic Properties of Generalized Lüroth Series," Acta Arithmetica, Vol. 74, No. 4, 1996, p 327.
4. K. Dajani and C. Kraaikamp, "On Approximation by Lü- roth Series," Journal de Théorie des Nombres de Bordeaux, V doi:10.5802/jtnb.172
5. K. J. Falconer, "Fractal Geometry, Mathematical Foundations and Applications," Wiley, Hoboken, 1990. 06, pp. 223-232.
6. K. J. Falconer, "Techniques in Fractal Geometry," Wiley, Hoboken, 1997.
7. J. Galambos, "Reprentations of Real Numbers by Infinite Series," Lecture Notes in Math, Springer, Berlin, 1976.
8. L. M. Shen and J. Wu, "On the Error-Sum Function of Lüroth Series," Mathematics Analysis and Applications, Vol. 329, No. 2, 2007, pp. 1440-1445. upported by the Hunan Educat partment Fund (11C671).
9. L. M. Shen, C. Ma and J. H. Zhang, "On the Error-Sum Function of Alternating Lüroth Series," Analysis in The- REFERENCES ory and Applications, Vol. 22, No. 3, 20 doi:10.1007/s10496-006-0223-x

10. T. Sálat and S. Znám, "On the Sums of Prime Powers," Acta Universitatis Palackianae Olomucensis of Mathematica, Vol. 21, 1968, pp. 21-25. - ernating Series Repres ta Arithmetica, Vol. 55 311-322.
11. 2. K. H. Indiekofer, A. Knopfmacher and J. Knopfmacher, "Alternating Balkema-Oppenheim Expansions of Real Numbers," Bulletin de la Société Mathématique, Vol.

CITATIONS

H. Jing and L. Shen, "Some Properties on the Error-Sum Function of Alternating Sylvester Series, doi: 10.4236/apm.2012.26070.

Index